VANADIUM IN BIOLOGICAL SYSTEMS

Vanadium in Biological Systems

Physiology and Biochemistry

Edited by

N. DENNIS CHASTEEN
Department of Chemistry, University of New Hampshire, U.S.A.

Kluwer Academic Publishers

Dordrecht / Boston / London

Library of Congress Cataloging-in-Publication Data

Vanadium in biological systems : physiology and biology / edited by N.
   Dennis Chasteen.
        p.   cm.
   Includes bibliographical references.
   ISBN 0-7923-0733-X (U.S. : alk. paper)
   1. Vanadium--Physiological effect.   I. Chasteen, N. Dennis.
QP535.V2V36   1990
574.19'214--dc20   90-4320

ISBN 0-7923-0733-X

Published by Kluwer Academic Publishers,
P.O. Box 17, 3300 AA Dordrecht, The Netherlands.

Kluwer Academic Publishers incorporates
the publishing programmes of
D. Reidel, Martinus Nijhoff, Dr. W. Junk and MTP Press.

Sold and distributed in the U.S.A. and Canada
by Kluwer Academic Publishers,
101 Philip Drive, Norwell, MA 02061, U.S.A.

In all other countries, sold and distributed
by Kluwer Academic Publishers Group,
P.O. Box 322, 3300 AH Dordrecht, The Netherlands.

Printed on acid-free paper

Printed in the Netherlands

Contents

Preface

Over the past several decades, vanadium has increasingly attracted the interest of biologists and chemists. The discovery by Henze in 1911 that certain marine ascidians accumulate the metal in their blood cells in unusually large quantities has done much to stimulate research on the role of vanadium in biology. In the intervening years, a large number of studies have been carried out to investigate the toxicity of vanadium in higher animals and to determine whether it is an essential trace element. That vanadium is a required element for a few selected organisms is now well established. Whether vanadium is essential for humans remains unclear although evidence increasingly suggests that it probably is.

The discovery by Cantley in 1977 that vanadate is a potent inhibitor of ATPases lead to numerous studies of the inhibitory and stimulatory effects of vanadium on phosphate metabolizing enzymes. As a consequence vanadates are now routinely used as probes to investigate the mechanisms of such enzymes. Our understanding of vanadium in these systems has been further enhanced by the work of Tracy and Gresser which has shown striking parallels between the chemistry of vanadates and phosphates and their biological compounds. The observation by Shechter and Karlish, and Dubyak and Kleinzeller in 1980 that vanadate is an insulin mimetic agent has opened a new area of research dealing with the hormonal effects of vanadium. The first vanadium containing enzyme, a bromoperoxidase from the marine alga *Ascophyllum nodosum*, was isolated in 1984 by Viltner. Shortly thereafter, a vanadium nitrogenase, which had been long suspected to exist, was unequivocally identified by Robson, Eady and coworkers in the nitrogen fixing bacterium *Azotobacter chroococcum*. As a consequence of these landmark studies, the number of papers published annually dealing with biological and biochemical aspects of vanadium has grown enormously in recent years (see graph).

I am pleased to present the first comprehensive book on the biochemistry of vanadium. Such a volume has been clearly overdue. In assembling the present work, I have tried to maintain a balance between the physiological, biochemical and chemical aspects of vanadium while addressing the important areas of current research. I have asked leading authorities to write in-depth accounts of ongoing work in the field while providing introductions for the nonspecialists.

Chapter I presents an overview of vanadium in biology including some toxicology. In Chapter II, the chemistry of vanadium in aqueous solution is reviewed along with some of the more recent developments. These two chapters should be required reading for anyone contemplating doing biological research involving vanadium. Chapter III reviews the recent literature on the metabolism and essen-

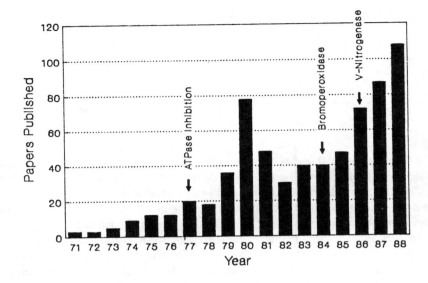

tiallity of vanadium as it relates to higher animals. In Chapter IV, the phosphate mimetic chemistry of vanadate and the inhibitory and stimulatory effects of vanadates on various enzyme systems are discussed. This chapter suggest a useful classification of phosphoryl transfer enzymes according to their interactions with vanadates. Chapters V and VI cover the first enzymes shown to require vanadium as a cofactor, namely the haloperoxidases and the nitrogenases, and discuss their structures and possible roles of the metal in catalysis. The insulin mimetic effects of vanadium and its utility in understanding the action of this hormone are reviewed in Chapter VII. The recent use of vanadium as an agent for the photosensitized cleavage of proteins, a technique that may have further applications in protein chemistry, is discussed in Chapter VIII. Chapter IX provides an update on vanadium in ascidians. Chapters X and XI present examples of the use of magnetic resonance methods (NMR, EPR, ESEEM, and ENDOR) in studies of vanadium containing biomolecules, demonstrating the utility of these techniques in obtaining structural information about the environment of the metal. These last two chapters provide good introductions to the literature for those planning to use magnetic resonance methods in their research.

It was my intent that *Vanadium in Biological Systems* should serve as a valuable reference work as well as an introduction to the chemistry and biochemistry of vanadium for those new to the field. Hopefully the volume will facilitate further interdisciplinary research and progress in this rapidly expanding field.

January, 1990   N. DENNIS CHASTEEN

I. Vanadium in the Biosphere

GAIL R. WILLSKY
Department of Biochemistry SUNY at Buffalo, School of Medicine and Biomedical Sciences, Buffalo, NY 14214, U.S.A.

I. Introduction to Vanadium

This opening chapter will strive to give an overview of the role of vanadium in the biosphere and to help place the subsequent chapters in the overall context of vanadium research. The amount of work on vanadium in biological systems has exploded since the late 1970's when Cantley and Josephson (1978) characterized the inhibition of the Na, K ATPase by vanadate and vanadium was shown to be a structural metal in algael bromoperoxidases (DeBoer and Wever, 1988) and bacterial nitrogenases (Smith *et al.*, 1988). Interest in vanadium was also intensified by the discovery of the insulin-like activity of vanadate in whole cell systems (Dubyak and Kleinzeller, 1980; Shechter and Karlish, 1980). The meticulous work of my colleagues who have assembled excellent reviews of various aspects of this field within the last few years will be heavily utilized and the reader is urged to consult these earlier articles. The potential role of vanadium in cellular regulation has been explored by Ramasarma and Crane (1981). The intriguing systems of the bivalve molluscs and tunicates, who accumulate large amounts of vanadium has been reviewed by Kustin and McLeod (1983). The biochemistry of vanadium has been surveyed with strong sections concerning the chemistry of vanadium by Boyd and Kustin (1984) and Chasteen (1983). The mechanisms of action of vanadium with an emphasis of the effects of vanadium on whole organ physiology was recently reviewed by Nechay (1984). In 1985 a FASEB Symposium was devoted to a discussion of the role of vanadium in Biology (Nechay *et al.*, 1986). A book on vanadium biochemistry complementing the present volume is in preparation (Kustin, 1990). This chapter will emphasize those areas of research not included in the above reviews. For a complete list of the effects of oxovanadium compounds on biological systems the reader is advised to consult those articles.

A. EARLY HISTORY

The credit for first recognizing a vanadium compound as a new metal goes to Nils Gabriel Sefstrom. In 1831 he named the oxide of an unusual constituent of iron ore, which formed beautiful multicolored compounds, vanadium after Vanadis the legendary Norse goddess of beauty. Subsequently the mineral was called

N. Dennis Chasteen (ed.), Vanadium in Biological Systems, 1–24.

vanadinite (Hudson, 1964). Vanadium is widely distributed in the world. The average concentration in the earth's crust is 100 ppm (Bertrand, 1950) which puts it ahead of copper, lead, zinc and tin in concentration in the crust. Due to its wide use in industrial processes as a catalyst, the concentration of vanadium is increasing in the atmosphere (Hudson, 1964).

Physiological effects of vanadium were studied long before its current renaissance. In 1876 John Priestly of Manchester reported on the physiological action of vanadium in frogs, pigeon, guinea pigs, rabbits, dogs and cats. Some of his conclusions are still valid today. For example, that there is a poisonous effect of the introduction of a soluble salt (a vanadate) into the system by the stomach or direct injection into veins. This report was followed by work concerning the toxicity of the various known vanadium forms (Larmuth, 1877). Vanadium was widely used for therapeutic purposes at the turn of the century in France for treatment of anaemia, tuberculosis, chronic rheumatism and diabetes (Lyonnett, 1899). In order to increase appetite, strength and weight, up to 5 mg/day of sodium metavanadate was given orally (Hudson, 1964).

B. INORGANIC CHEMISTRY

The following sections describe the inorganic chemistry of vanadium which is relevant to experiments in biological systems. The major points are the pH and concentration dependent equilibria and the various geometries associated with both vanadate [vanadium in the + 5 oxidation state, V(V)] and vanadyl [vanadium in the + 4 oxidation state, V(IV)] structures. The latter point is important because of increasing evidence that both vanadate and vanadyl have similar effects on some biological systems. If these effects are mediated via one site on a single enzyme then it would appear that the two oxovanadium forms must be able to assume similar geometries. However, to date the detailed interactions of both vanadate and vanadyl compounds at the same site of any protein have not been extensively characterized. The information in this section is mostly summarized from the reviews of Chasteen (1983) and of Boyd and Kustin (1984), and the textbook of Cotton and Wilkenson (1980). If further information is needed, the reader is referred to these sources and Chapter II of this volume which details the coordination chemistry of vanadium in aqueous solution.

1. Vanadium (V). The speciation of vanadate, V(V), as a function of pH is given in Figure 1 for high and low concentrations of vanadate. The form of vanadate which most closely mimics phosphate, VO_4^{3-} is only present at extremely high pH values at which biological systems are mostly inactive. At concentrations of vanadate below 10 μM, which is the physiological range, the monomeric form of vanadate can be found, with the tetravalently coordinated $H_2VO_4^-$ and HVO_4^{2-} being the predominant species at pH values between 6 and 8. Diagram (i) shows the structure of the anhydrous KVO_3 in which chains of VO_4 tetrahedra share corners. The VO_4 structure is very similar to that of PO_4. The vanadate ion $HV_4O_{12}^{3-}$ is commonly found in mM vanadate solutions at physiological pH

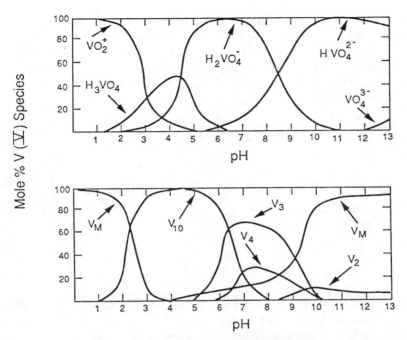

Fig. 1. The speciation of V(V) as a function of pH. A) total vanadium concentration is equal to 10 μM. B) Total vanadium concentration is equal to 10 mM. Vm is the sum of all species containing 1 V(V) atom, Vn is the sum of all species containing n V(V) atoms, n = 2,3,4,10 (redrawn from Boyd and Kustin, 1983). V_3 is now known not to exist in solution (see Figure 2, chapter X).

values. Its structure is a cyclic anion with four VO_4 units formed by linking corners, similar to tetrametaphosphate. The structure of the hydrated $KVO_3 \cdot H_2O$ (ii) shows how vanadate can exist as linked VO_5 polyhedra in a trigonal bipyramidal structure. Note that at higher concentrations of vanadate, which could occur naturally if vanadate were concentrated in vesicular structures, there would be some decavanadate [V(V), $V_{10}O_{28}^{6-}$] present at pH ranges between 6 and 8. Decavanadate is the major oxovanadate species present at pH 5.5, which is the pH of many vesicles in eucaryotic cells, at concentrations above 0.5 mM monovanadium units. The structure of decavanadate is a rigidly defined polymer (iii) in which vanadium atoms exist in three different electronic configurations (Howarth and Jarrold,1984). The vanadate geometries will vary if the vanadate is complexed with other structures. Vanadate is known to bind to EDTA (Kustin and Toppen, 1973) citrate, succinate, glycols and catechols (Boyd and Kustin, 1984). Vanadate solutions are usually colorless, except for the polymerized decavanadate which forms an orange-yellow color.

The effect of vanadate on phosphohydrolases, which have a stable phosphoprotein intermediate in their catalytic cycle, is believed to occur due to similarities between the chemistries of vanadates [V(V)] and the phosphates. Vanadate can readily acquire a stable five-coordinate trigonal bipyramidal geometry resembling the transition state of phosphate. V-O bond lengths in vanadates are around 0.17 nm compared to 0.152 for the P-O bond in orthophosphate. Vanadic acid H_3VO_4 (pKa = 3.5, 7.8, 12.5) is a weaker acid than phosphoric acid H_3PO_4 (pKa = 1.7, 6.5, 12.1). The phosphate analog aspects of vanadium chemistry is further explored by Gresser and Tracey in Chapter IV of this volume.

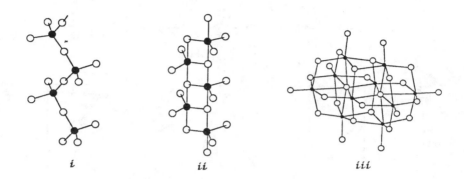

i   *ii*   *iii*

2. Vanadium (IV). Vanadate can easily be reduced to vanadyl by glutathione, catecols and other cellular constituents. The coordination chemistry of vanadyl (VO^{2+}), has been reviewed in detail by Chasteen (1981). Five and six coordinate complexes are formed in which the short V-O bond length of 0.16 nM is preserved. Complexes usually approach one of three idealized geometries square bipyrimidal (iv), square pyrimidal (v) or trigonal bipyramidal (vi). Note the similarity of structure (vi) to the trigonal bipyramidal structure vanadate can assume (ii). This geometry for vanadyl species is usually found only in protein complexes. As vanadyl, VO^{2+}, the vanadium ion will behave as a simple divalent metal and will compete favorably with Ca^{++}, Mn^{++}, and Zn^{++}, etc., for ligand binding sites. The ability to bind in different metal binding sites is greatly aided by its flexible coordination geometry. Vanadyl solutions are usually blue.

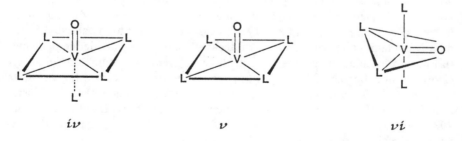

iv   *v*   *vi*

The speciation diagram for low physiological concentrations of vanadyl, V(IV) as a function of pH is given in Figure 2. Note that distribution diagrams for VO^{2+} are incomplete. The vanadyl species present as a function of pH at 2×10^{-2} M VO^{2+} has been calculated by Labonette (1988). Above pH 2.3 vanadyl(IV) ion

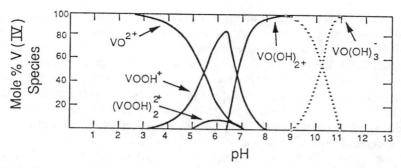

Fig. 2. The speciation of V(V) as a function of pH. Total vanadium concentration is equal to $10 \, \mu M$ (redrawn from Chasteen, 1983).

tends to undergo air oxidation to form vanadates. Hydrolyzed species of VO^{2+} are very prone toward air oxidation. However, when the VO^{2+} is chelated, oxidation is considerably retarded. VO^{2+} forms extremely strong complexes with proteins (apparent association constants of $10^9 \, M^{-1}$) and also complexes with ligands such as citrate, ATP, pyrophosphate, catechols and free amino acids. It is unlikely that VO^{2+} exists free in biological systems at physiological pH, except perhaps in acidic vesicles, due to its tendency to bind to other compounds and oxidize to vanadate in the free form.

II. Vanadium in Biological Systems

The essentiality, distribution and toxicology of vanadium are all areas of ongoing research at the present time. Vanadium is an ultratrace metal, present in mammalian tissues at concentrations below $1 \, \mu M$ (Nechay, 1984). The ascidians (cf. Chapter IX by Michibata and Sakurai) have extremely high concentrations of vanadium (approaching 1M) in specialized cells, vanadocytes which concentrate vanadium. Plants also have measurable amounts of vanadium (Nechay, 1984). The essentiality of the metal in mammalian systems is an area of controversy. The pharmacological effects of vanadium add to the difficulty in interpreting experiments designed to determine its essentiality. The current state of this field is summarized in Chapter III by Nielsen and Uthus. Vanadium is the structural metal in an alternative bacterial nitrogenase (cf. Chapter VI) and algal haloperoxidase (cf. Chapter V). Due to the increasing amounts of vanadium in the atmosphere as a result of its use in industrial processes (Hudson, 1964), the use of vanadium in dental implants (Blumenthal and Cosma, 1989) and the pharmacological effects of vanadium, the toxicology of vanadium has become an area of great interest.

The toxicity of Vanadium compounds has been studied since the late 1800's. Human vanadium toxicity has been predominantly associated with industrial processes (Hudson, 1964). The general topic of vanadium toxicity has been discussed by Kubena and Phillips (in Nechay *et al.*, 1986). Absorption of vanadium may be gastrointestinal, respiratory of percutaneous and it is mainly excreted in the urine. Toxicity increases as the valency increases, with pentavalent vanadium being the most toxic (Nechay *et al.*, 1986). However, oral administration of vanadium is not very toxic in humans, as seen from the widespread medicinal use of vanadium at the turn of the century. The major toxic effects of vanadium in humans has been observed as a result of breathing in vanadium dust in industrial processes. Using the rat model, a detailed description of the effect of inhalation of bismuth orthovanadate on pulmonary tissue has been reported (Lee and Gillies, 1986). A compartmental model for whole-body vanadium metabolism was hypothesized from studies in sheep fed vanadium in the diet (Patterson *et al.*, 1986). The effect of oral administration of vanadate in drinking water was studied in rats (Domingo *et al.*, 1985). Vanadium accumulated predominantly in the kidneys and the spleen over a three month period using doses of 0-50 ppm. The toxicity of subcutaneous injection of 20 mg vanadate/kg was done in mice (Al Bayati, 1982). Acidification of the solution did reduce toxicity. Vanadate treated animals showed a tendency to increase their liver and spleen weights, exhibited severe necrosis in lymphoid tissues, pulmonary hemorrhage and renal acute tubular necrosis. Recent work has shown that chelating agents protect against the toxic effects of vanadium in mice (Jones and Basinger, 1983; Domingo *et al.*, 1986). The effect of chelating agents on vanadium distribution in rats and uptake into erythrocytes has also been studied (Hansen *et al.*, 1982). Both Desferal and Ca-Na$_3$-diethylene-triaminepentaacetic acid (at 100 μM) raised faecal and urinary excretion in the rat. The chelators also reduced the amount of vanadium bound to the red cells.

The ability to assay for the various oxovanadium species in biological experiments is crucial to understanding the role of this metal in biology. The uses of atomic absorption spectrometry, neutron activation analysis, inductively coupled plasma, dc arc emission, dc plasma atomic emission spectroscopy, mass spectrometry, spectrophotometry, energy dispersive x-ray fluorescence and electron paramagnetic resonance spectroscopy to determine vanadium content in biological samples are described in Martin and Chasteen (1988). However, it is often useful to be able to specifically measure vanadium concentrations in a given oxidation state by chemical assay. N-benzoyl-N-phenylhydroxylamine is a specific reagent for vanadium(V) and can be used to assay V(V) in biological samples (Priyadarshini and Tandon, 1961). A quantitative enzyme assay procedure for determining the amount of V(V) and V(IV) in biological samples has been developed based on vanadium inhibition of alkaline or acid phosphatases (Crans *et al.*, 1989).

Some useful properties of the oxovanadium species of importance to biological systems are shown in Table I. Note that decavanadate will form instantaneously if the pH and concentration is correct. Once formed decavanadate persists for hours when it is diluted to concentration and pH conditions which do not favor its formation. Also, metavanadate will form immediately if the pH of an ortho-

Table I. Properties of principal oxovanadium compounds active in biological systems.

Oxyvanadium species (1 mM)	Structure	Vanadium oxidation state	Solution pH (in Na$^+$ or SO$_4{}^{3-}$)	^{51}V-NMR chemical shift (ppm)[a]	ESR spectra
Orthovanadate	VO$_4{}^{3-}$	V	>12	−541[b]	No
Metavanadate		V	8	No	
	Monomeric			−555[c]	
	Dimeric			−570[c]	
	Tetrameric			−575[ce]	
	Pentameric			−582[c]	
Decavanadate	V$_{10}$O$_{28}{}^{6-}$	V	4	−423, −498, −517[d]	No
Vanadyl	VO^{2+}	IV	2	No	Yes

[a] The chemical shift is given in reference to a VOCl$_3$ standard (0ppm) and will vary with changes in buffers. Note that for decavanadate 3 resonances are seen for one structure, while for metavanadate multiple resonances arise from an equilibria of different vanadate species at pH 8.0.
[b] From Heath and Howarth (1981).
[c] From Tracy *et al.* (1988).
[d] From Howarth and Jarrold (1984).
[e] Major vanadate resonance.

vanadate solution is lowered. At physiological concentrations ($<100\ \mu$M) meta-vanadate will exist as the monomeric species at neutral pH (HVO$_4{}^{2-}$, H$_2$VO$_4{}^-$ cf. Figure 1A).

A. CELLFREE SYSTEMS

The most straightforward experiments to interpret concerning the effects of oxo-vanadium compounds on biological systems are carried out in cell-free systems. In these systems cellular interconversion of oxovanadium compounds is not as serious a problem as it is in whole cell experiments. However, one must be sure to consider the vanadium binding properties of all components of the buffer system, when postulating or calculating the concentration of free oxovanadium compounds.

1. Inhibition

The inhibitory effects of vanadium on the Na,K ATPase *in vitro* first reawakened interest in interactions of vanadium with metabolic processes. Table II gives a list of enzymes, their source and the reported apparent vanadate Ki for these systems. The mechanism of vanadate inhibition for most of these systems is well under-stood for enzymes having a phosphoprotein intermediate in their catalytic cycle. These enzymes appear to have a phosphate linked to the protein through an aspartate residue. The vanadate, which has trigonal bipyrimidal geometry, binds to the active site of the enzyme and acts as a transition state analogue. However,

Table II. Vanadium inhibition constants for various enzymes[a].

Enzyme	Source	Form of Vanadium[b]	$K_I(M)$[c]	Reference
Adenylate kinase	Skeletal muscle	DecaV(V)	$<1 \times 10^{-6}$	DeMaster and Mitchell (1973)
Acid phosphatase	Human liver	V(V)	2×10^{-7}	VanEtten et al. (1974)
Acid phosphatase	Wheat germ	V(V)	6.7×10^{-6}	VanEtten et al. (1974)
Alkaline phosphatase	E. coli	V(V)	4×10^{-7}	Lopez et al. (1976)
Alkaline phosphatase	Human kidney, liver, intestine	V(V)	$(0.5 - 0.9) \times 10^{-6}$	Seargeant and Stinson (1979)
ATPase	Mitochondria	V(V)[b]	1.5×10^{-3}	Schwartz et al. (1980)
Transducin GTPase	Bovine retina	DecaV(V)	2.5×10^{-4}	Kanaho et al. (1980)
Ca^{2+}, Mg^{2+}-ATPase	Cardiac sarcoplasmic reticulum	V(V)	$(5 - 50) \times 10^{-6}$	Hagenmeyer et al. (1980) Schwartz et al. (1980)
Ca^{2+} uptake	Sarcoplasmic reticulum	V(V)[b]	2×10^{-5}	Schwartz et al. (1980)
Ca^{2+}, Mg^{2+}-ATPase	Red cell	V(V)	$0.8 \times x10^{-6}$, 1.5×10^{-6}	Bond and Hudgins (1980) Barrarubin et al. (1980)
Ca^{2+}-ATPase	Squid axon	V(V)	$7 \times x10^{-6}$	Dipolo and Beauge (1981)
Dynein ATPase	Sea urchin sperm	V(V)[b]	10^{-6}	Gibbons et al. (1978)
Dynein ATPase	Tetrahymena	V(V)	$1 \times 10^{-8} - 5 \times 10^{-6}$	Himizu (1981), Kobayashi et al. (1978)
Mg^{2+}-ATPase	Yeast plasma membrane	V(V)	$(0.14 - 0.20) \times 10^{-6}$, 1×10^{-6}	Borst-Pauwels et al. (1981)
Mg^{2+}-ATPase	Neurospora crassa	V(V)	$(0.45 - 1.0) \times 10^{-6}$	Bowman and Slayman (1979)
Myosin ATPase	Muscle	V(V)	$<10^{-6}$	Goodno (1979) Goodno and Taylor (1982)
Na^+, K^+-ATPase	Kidney	V(IV)	8×10^{-6}	Schwartz et al. (1980)
Na^+, K^+-ATPase	Heart atria and ventricle	V(V)	0.6×10^{-6}	Borchard et al. (1979)
Na^+, K^+-ATPase	Skeletal muscle membrane	V(V)	1×10^{-6}	Byskocil et al. (1981)
Na^+, K^+-ATPase	Sarcolemma	V(V)	1×10^{-6}	Quist and Hokin (1978)
K^+-p-nitrophenol phosphatase	Skeletal muscle membrane	V(IV) & V(V)	4×10^{-8} 4×10^{-8}	Byskocil et al. (1981)

Table II (continued)

Enzyme	Source	Form of Vanadium[b]	$K_I(M)^c$	Reference
Phosphofructo-kinase 1	Sheep heart	DecaV(V)	4.5×10^{-8}, 5.5×10^{-7}	Choate and Mansour (1979)
6-phosphofructo kinase 2	Rat liver	DecaV(V)	2.5×10^{-5} M	Kantz *et al.* (1986)
DPG-dependent phosphoglycerate mutase	Rabbit muscle	V(V)	5×10^{-6}	Carreras *et al.* (1980)
Ribonuclease	Bovine pancreas	V(IV)	6×10^{-5}	Lindquist *et al.* (1973)
Ribonuclease	Bovine pancreas	V(IV): uridine	1×10^{-5}	Lindquist *et al.* (1973)
Ribonuclease	Bovine pancreas	V(V): uridine	1.2×10^{-5}	Lindquist *et al.* (1973)
Staphylococcal nuclease	Staphylococcus	$Ca^{2+}/VO^{2+}/$ thymidine 3-phsophate	1.7×10^{-8}	Tucker *et al.*(1979)

[a] Modified from Chasteen (1983).
[b] In most instances the chemical identity of the inhibitor is not known with certainty. It is generally believed that most vanadate(V) inhibitors correspond to one of the protonated forms of ortho-vanadate(V). In some cases decavanadate (DecaV) is known to be the inhibitory form of vanadate.
[c] Inhibitor dissociation constant. Some values listed correspond in the original work to the concentration of inhibitor which reduced the observed enzyme catalyzed rate by 50%. In these cases kinetic analyses using the standard equations for enzyme inhibition were not employed.

it is not clear that inhibition by the well ordered polymeric decavanadate occurs via the same mechanism. The inhibition of the Na,K ATPase by vanadyl V(IV) (North and Post, 1984) is much weaker than that found by vanadate, V(V).

2. Stimulation

The mechanism by which vanadate stimulates biological processes is much less well understood than the mechanism of inhibition by vanadate. One problem has been the lack of data demonstrating which oxovanadium form of vanadate is the active species. At the current time the chemical mechanisms of stimulatory actions of oxovanadium compounds remain elusive.

a. Adenylate Cyclase. This enzyme is one of the first to have been shown to be stimulated by vanadate in the + 5 oxidation state. The stimulation by vanadate differs from that of fluoride (Johnson, 1982). Vanadate has been hypothesized to act via guanine nucleotide regulatory protein (Krawietz *et al.*, 1982).

b. Vanadate-Dependent NADH Oxidation. Vanadate-stimulates NADH oxidation activity in the absence of added protein. Stimulation by protein of vanadate-dependent NADH oxidation by plasma membranes was reported by Ramasarma *et al.* in 1981. This plasma membrane activity has been studied in plants (Briskin *et al..*, 1985), the yeast *S. cerevisiae* (Willsky, unpublished observations), cardiac cells (Erdman *et al.*, 1979) erythrocytes (Vijaya *et al.*, 1984) and rat liver (Coulombe *et al.*, 1987). The plasma membrane-mediated, vanadate stimulated NADH oxidation activity is only observed in phosphate buffers. NADPH will substitute for NADH. Superoxide production is involved because the reaction is inhibited by superoxide dismutase,a superoxide radical scavenger, and is stimulated by paraquat, a superoxide producer. The reaction does not occur anaerobically.

The following reaction scheme has been proposed by Darr and Fridovich (1984)

$$V(V) + O_2^- \rightleftarrows V(IV)\text{-}OO$$
$$V(IV)\text{-}OO + NADH \rightleftarrows V(IV)\text{-}OOH + NAD^{\cdot}$$
$$V(IV)\text{-}OOH + H^+ \rightleftarrows V(V) + H_2O_2$$
$$NAD^{\cdot} + O_2 \rightleftarrows NAD^+ + O_2^-$$

Note that in the presence of the autoxidation of NADH this chain reaction can continue in the absence of protein. The net result of these reactions is the formation of H_2O_2. The plasma membrane stimulated vanadate-dependent NADH oxidation activity is hypothesized to be initiated by the autoxidation of NADH, which results in the formation of the first superoxide radical (O_2^-).

The purpose of the plasma membrane-stimulated, vanadate-requiring NADH oxidation activity remains elusive. The function of the membrane portion of the reaction appears to be the production of superoxide, which is potentially toxic to the cell, while the continuation of the chain causes the production of hydrogen peroxide, another toxic compound. It has been suggested that the reaction is involved in the generation of heat and/or vanadium toxicity. However, we have shown that the facultative anaerobic yeast, *S. cerevisiae*, is equally sensitive to vanadate when grown aerobically or anaerobically, which implies that no oxidative processes of any type are involved in vanadate toxicity (Minasi *et al.*, in press).

Vanadyl has also been shown to stimulate NADH oxidation activity in a two phase process in which the first, rapid phase involves the production of vanadate as shown in the following reaction mechanism (Liochev and Fridovich, 1986):

$$V(IV) + H_2O_2 \rightleftarrows V(V) + OH^- + OH^{\cdot}$$
$$OH^{\cdot} + NADH \rightleftarrows NAD^{\cdot} + H_2O$$
$$NAD^{\cdot} + O_2 \rightleftarrows NAD^+ + O_2^-$$
$$OH^{\cdot} + H_2O_2 \rightleftarrows OH^- + 2H^+ + O_2$$

c. Protein Phosphorylation. The effect of vanadium compounds on protein phosphorylation is not well understood and only a small fraction of the experiments in this field have been done in cell free systems. In isolated cell membranes from A-431 cells it has been reported that μmolar concentrations of vanadate inhibit the membrane-dependent dephosphorylation of histones containing phospho-tyrosine, but not those containing phosphoserine or phosphothreonine (Swarup *et al.*, 1983). Vanadate (in μmolar quantities) also increased tyrosine phos-phoryation in a membrane fraction from a human lymphoblastoic cell (Earp *et al.*, 1983). In experiments using insulin-receptor enriched fractions obtained from microsomal membranes of human placenta and rat liver, vanadate activation of the autophosphorylation of the insulin receptor was observed. The autophos-phorylation occurred in a dose dependent manner, showing a half maximal response at 30 μM (Gherzi *et al.*, 1988). In experiments using cellular fractions from rat brain cortex, vanadate and vanadyl ions stimulated total phosphorylation of proteins from synaptic membranes and to a lesser extent from mitochondria (Krivanek, 1988). Vanadate has been shown to rapidly esterify the hydroxyl groups of aromatic rings to yield a phenyl vanadate. The vanadate esterification proceeds with an equilibration constant much larger than that of the correspond-ing phosphate esterification reactions (Tracy and Gresser, 1986).

B. ORGANS AFFECTED BY EXOGENOUS VANADIUM

Exogenously added oxovanadium compounds have been shown to have many physiological effects on various organ systems. At the current time it is not possible to differentiate primary from secondary effects of vanadium. In addition, the form of oxovanadium which is affecting each system has not yet been deter-mined due to the pH, concentration and oxidation-reduction changes that can occur as the oxovanadium compounds move through the body. Throughout the body the most consistent response is one of vasoconstriction. There have been some studies concerning the effect of vanadate on eye, ear, liver and brain (Nechay, 1984; Nechay *et al.*, 1986). However, the bulk of the work consists of studies utilizing the kidney and heart.

a. Heart. The effects of vanadate on the heart have been reviewed (Akera *et al.*, 1983; Erdmann *et al.*, 1980). The addition of oxovanadium compounds to cardiac muscle has been shown to have both positive and negative inotropic effects, depending upon the animal species studied and the amount of drug used in the study (Inciate *et al.*, 1980; Takeda *et al.*, 1980). Alterations in cardiac function by vanadate have not been linked to changes in K or Na transport in the heart implying that interaction with the Na,K ATPase is not the primary action of oxovanadium in the heart. The addition of calcium channel blockers, such as verapamil, attenuate increases in pressure and diminish increases in cardiac output caused by vanadate (Sundet *et al.*, 1984). Alterations in calcium transport have also been reported after vanadate addition to cardiac tissue as described in section IICe below. Vanadate does stimulate rat heart protein kinase c, which could cause changes in calcium mobilization through action of the inositol phos-phate system (Catalan *et al.*, 1982).

b. Kidney. Vanadate accumulates to greatest levels in the kidneys and has a strong consistent vasoconstriction effect in that organ (Nechay, 1984). Vanadate causes strong naturesis and diuresis in various animals. The vasoconstriction produced in rat (Day *et al.*, 1980) is much less intense than that in the more sensitive dog. Alterations of calcium concentration in kidney perfusates have been shown to decrease the vasoconstrictive effect of vanadate (Benabe *et al.*, 1984). The effects of vanadate and other diuretics are additive. In isolated systems, vanadate is more potent when present in the lumen compared to the peritubular bath of isolated perfused renal tubules (Day *et al.*, 1980). Vanadate also inhibits transtubule secretion of p-aminohyppuric acid (Edwards and Grantham, 1983) and inhibits the action of vasopressin in the collecting tubule (Stefan *et al.*, 1981). The possible inhibitory interactions of vanadate on Na,K ATPase have been considered as a mechanism of action of vanadate in kidney, since K and Na fluxes are affected by vanadate (Higashine *et al.*, 1983; Churchill and Churchill, 1980). Vanadate also reduces renal renin secretion in rat kidney slices (Churchill and Churchill, 1980).

C. PHYSIOLOGICAL EFFECTS OF VANADIUM

Exogenous vanadium has been shown to affect various physiological processes. The addition of oxovanadium to whole animals or cells has a variety of physiological effects which will be discussed here as vanadium-induced alterations of normal physiological function and vanadium interactions with disease states.

1. Effects of Vanadium on Cellular Processes

Vanadate addition has been shown to alter various cellular processes. Vanadate inhibited the ATP-dependent degradation of proteins in reticulocytes (Tanaka *et al.*, 1984). Vanadate has also been shown to affect lipid metabolism. Administration of vanadate intraperitoneally to rats has been shown to increase brain cholesterol levels and the ratio of cholesterol/phospholipid. Vanadate addition had a similar, though reduced, effect on lipids in *in vitro* experiments with cultured cells (Catalan, 1987). Vanadate has also been reported to inhibit the rate limiting step in cholesterol biosynthesis, the synthesis of mevalonate, which is catalyzed by HMG CoA (3-hydroxyl-3-methylglutaryl coenzyme A) reductase (Menon *et al.*, 1980).

a. Insulin-Mimetic Activity. In 1980, Shechter and Karlish, and Dubyak and Kleinzeller concurrently demonstrated that the addition of vanadate to isolated adipocytes increased glucose transport and metabolism. In 1985, vanadate was first reported to have insulin-mimetic effects in whole animal studies (Heyliger *et al.*, 1985). Oral administration of vanadate to rats with streptozotocin-induced diabetes, lowered blood glucose values and returned abnormal cardiac function (measured in an isolated working heart system) to near normal levels. A detailed analysis of the insulin-mimetic activity of vanadate is given by Shechter *et al.*, in Chapter VII of this volume. Vanadate treatment has recently been shown to have transcriptional effects, inducing amylase mRNA synthesis in diabetic rat pancreas (Johnson *et al.*, 1990).

b. Cell Proliferation, Differentiation and Phosphorylation/Dephosphorylation Activity.
The stimulatory actions of vanadium on protein phosphorylation may be
related to its hormone-mimetic activities (see Chapter VII), its potential role in the
regulation of cell proliferation and differentiation (Nechay *et al.*, 1986), and its
anticarcinogenic properties (see below). This section will be divided into work
which focuses on cell proliferation and differentiation and that which focuses on
phosphorylation/dephosphorylation systems. Both areas were discussed in the
1985 FASEB meeting (Nechay *et al.*, 1986).

i. Cell Proliferation and Differentiation. Vanadate addition to culture medium has
produced numerous stimulatory effects on cell proliferation and differentiation.
Quiescent human fibroblasts treated with 4 μm vanadate were stimulated to divide
and DNA synthesis increased. However, at 210 μM vanadate in the culture
medium no stimulation of DNA synthesis was found (Carpenter, 1980). Vanadate
was also found to replace interleukin 3 as a growth factor in a mouse cell line (Tojo
et al., 1987). In friend erythroleukemia cells vanadate blocked differentiation
without affecting cell viability (English *et al.*, 1983). In human lymphocytes
vanadate had an inhibitory effect during the first 3 days of cell culture, when both
differentiation and proliferation take place, but it enhanced DNA synthesis, acting
as a co-mitogen in subsequent days of culture (Marini *et al.*, 1987). Vanadate
added earlier to the culture medium had been found to be mitogenic for a subpopulation of thymus cells, but not for splenocytes and vanadate inhibited the
conA mitogenic response (Ramanadham and Kern, 1983). Vanadate has been
shown to stimulate bone cell proliferation and collagen synthesis at 5-15 μM (Lau
et al., 1988). Vanadate, *in vitro*, potentiated the estrogen stimulation of proliferation of a Mouse Leydig cell line (Sata *et al.*, 1987).

ii. Phosphorylation/Dephosphorylation. The effects of vanadate treatment on phosphoprotein levels have been examined in various systems. A stimulatory effect of
vanadate treatment on protein tyrosine phosphorylation in fibroblasts has been
reported (Klarland *et al.*, 1988). Biochemical assay for phosphotyrosine residues
was used to show that vanadate treatment stimulated phosphotyrosine production
in Mouse lens cell (Gentleman *et al.*, 1987). The occurrence of protein phosphorylation has been studied in conjunction with the insulin-like activity of
vanadate. Phosphorylation of the insulin receptor in vanadate-treated adipocytes
(Tamura *et al.*, 1983) was reported. However, other workers using an anti-
phosphotyrosine antibody were unable to see an increase in tyrosine protein
phosphorylation in the adipocyte system (Mooney *et al.*, 1989). Most recently the
effect on oncogenes of incubation with vanadate has been reported. Vanadate
stimulated phosphorylation of the *fms* (Tamura *et al.*, 1986), and *src* (Ryer and
Gorman, 1987; Collet and Belzer, 1987) proteins. These oncogenes are believed
to be phosphorylated by protein kinase C whose stimulation by the degradation
of membrane phosphoinositol depends on intracellular calcium mobilization.
Vanadate has also been shown to induce inositol phosphate formation and inhibit
inositol phosphate degradation in CCL39 cell lines (Paris and Poussegur, 1987).

c. Cellular and Intracellular Motility. The motility of sperm, cilia and chromosomes have been reported to be inhibited by vanadate (Ramasandra and Crane, 1981). This effect may be related to the inhibition of dynein and muscle ATPases by vanadate (Nechay, 1984; Nechany *et al.*, 1986). Intracellular movement of vesicles may be stimulated by vanadate. It has been demonstrated that vanadate addition to yeast growth medium stimulated the accumulation of the fluorescent dye lucifer yellow in yeast, which is used as a measure of endocytosis (Reisman, 1985).

d. Transport Across the Plasma Membrane. The effects of vanadate-treatment on transport into cells has been extensively studied in a few cell types. Vanadate-treated rat adipocytes have been used to study transport in many experiments designed to characterize the insulin-mimetic activity of vanadate (cf. Chapter VII). The effect of vanadate treatment on glucose and cation transport in isolated adipose tissue and skeletal muscle has been studied (Clausen *et al.*, 1981). In fat pads 0.5 to 5 mM vanadate stimulated glucose uptake and efflux of 3-0-methyl glucose and increased calcium washout from preloaded fat pads. In muscle 3-0-methyl glucose efflux and calcium washout were stimulated, while in other muscle cells vanadate treatment increased K while decreasing Na. The stimulatory effect of vanadate on glucose transport in this case appears to be associated with or mediated by a rise in cytoplasmic calcium level (See calcium movement section below).

Vanadate has been shown to stimulate ion movement across various portions of the gastrointestinal tract. In rabbit colonic epithelia μM amounts of vanadate on the serosal side caused the stimulation of chloride secretion, without affecting sodium transport (Hatch *et al.*, 1983). In rat jejunum vanadate appears to have a concentration-dependent affect on transport. At 0.1 mM it increased influx and efflux of alanine across the mucosal border, but at higher concentrations (1 to 10 mM) it decreased mucosal to serosal flux and influx of alanine concurrently with reducing Na,K ATPase activity in basolateral membranes (Hajjar *et al.*, 1987). It has been suggested that the low concentration effects result from vanadyl stimulation of adenylate cyclase (see above).

Vanadate affects on transport in the red blood cell have also been monitored. Calcium transport into red cells was stimulated by 0.5 mM vanadate, accompanied by a large efflux of K^+. It is hypothesized that in these cells vanadate moves inside the cell and the increased calcium accumulation is due to decreased activity of the Ca^{++} ATPase which pumps calcium out of the cell (Varecka and Carafoli, 1982). However, using measurements of K permeability and single-channel currents, it was shown that vanadate stimulated K permeability by increasing the probability of channel openings (Furhmann *et al.*, 1987). Vanadate inhibited calcium efflux from squid axon (Nordmann and Zyzek, 1982). In isolated rat neurohypopyses calcium efflux was inhibited while Na,Ca^{++} exchange was not affected (DiPolo *et al.*, 1979).

e. Intracellular Ca^{++} Movement. Vanadate interference with intracellular calcium movement would link the hormonal affects of vanadate to interactions of protein kinase c. Although the subject of hypothesis, little direct work has been done on the effect of vanadium on intracellular calcium movement. ATP dependent calcium uptake by microsomal fractions of rat salivary gland was inhibited by vanadate (Kanagasuntheran and Theo, 1982). Alterations in calcium movement have been associated with the inotropic effects of vanadate (Akera et al., 1983). The increased vanadate induced force of contraction of isolated rat atrial muscle is associated with increases of the Mn sensitive calcium pool, while vanadate lowering of the force of contraction of guinea pig atrial muscle has been associated with inhibition of the closing of calcium channels. Vanadate stimulation of glucose metabolism in adipocytes and skeletal muscle is associated with a rise in cytoplasmic calcium concentration (Erdmann, 1980).

f. Oxidation-Reduction Processes. From vanadate chemistry it is expected that vanadium ions would affect oxidation reduction processes in cells. However, there has not been extensive work directed at identifying redox processes in cells that may be affected by vanadate. Whether vanadate-dependent NADH oxidation activity is related to any cellular phenomena is not not known. The ability of *S. cerevisiae* to reduce vanadate to vanadyl has been shown to be increased in vanadate resistant mutants, implying that this pathway may aid in lessening the toxic effects of vanadate (Willsky et al., 1985). Vanadate is taken up by cells and rapidly reduced to vanadyl (Degani, 1981; Willsky et al., 1984). This reduction is mediated by glutathione and cellular catechols (Macara et al., 1980; Cantley and Aisen, 1979). The formation of an intermediate vanadate-thioester has been proposed as part of a preequilibrium step in glutathione reduction of vanadate to vanadyl (Legrum, 1986). Administration of vanadate to mice depresses the oxidation rate of formate to CO_2, while *in vitro* in cytosolic liver fractions vanadate inhibited the enzymatic transfer of formate to tetrahydrofolic acid (Bruch et al., 1987). In mice vanadate inhibited the oxidative demethylation of substrates of the cytochrome P-450 dependent monooxygenase system. The effect was reduced by pretreatment of the animals with ascorbic acid and vanadyl sulphate did not produce as marked an effect as orthovanadate (Heide et al., 1983). It appears that in this system vanadate is the more potent inhibitor and reduction to vanadyl by ascorbate lessens the effect of vanadate. Effects of vanadate on one carbon metabolism and cytochrome P-450-dependent processes certainly implicate vanadate as a modulator of oxidative-reduction reactions in the cell.

2. Effects of Vanadium on Disease Processes (Excluding Toxicity)

a. Diabetes. Oral treatment with vanadate of diabetic rats obtained by streptozotocin treatment has been shown to alleviate the symptoms of diabetes. This effect is almost certainly related to the insulin-mimetic activity of vanadate demonstrated in isolated cell systems. These findings are discussed in Chapter VII. The heart, kidney and eye are the major target organs in diabetic animals. Oral vanadate and vanadyl treatment has been shown to restore weakened cardiac function to diabetic rat hearts using the working heart preparation (Ramanadham, *et al.*, 1989); while oral vanadate treatment has been shown to alleviate the diabetic renal hypertrophy and sorbitol accumulation (Pochal *et al.*, 1989).

b. Carcinogenesis. In mice there has been shown to be a 50% reduction in the ability to induce mammary carcinomas with N-methyl nitroso urea for animals fed vanadyl sulfate (Thompson *et al.*, 1984). The interactions of vanadium with phosphorylation/dephosphorylation processes, especially involving known oncogene products (cf. section C, 1b), will undoubtedly be related to its effects on carcinogenesis.

c. Depressive Illnesses. The levels of vanadium in depressive illness have been examined (Naylor *et al.*, 1987). Vanadium concentrations in hair, whole blood and serum were greater in depressed patients compared to vanadate levels after recovery in the same patients. Renal clearance was lower and mean serum vanadium concentation higher using two different populations of depressed patients compared to controls. Finally, there was a strong negative correlation between Na,K ATPase activity in erythrocytes and serum vanadium concentrations, supporting the hypothesis that changes in tissue vanadium concentration may explain the changes in sodium transport which occur in depressive psychosis. Patients taking lithium for depressive illness were found to have lower serum vanadium, and cobalt, with elevated serum aluminum (Campbell *et al.*, 1988).

d. Hypertension. Oxovanadium compounds cause vasoconstriction and increased arterial blood pressure. Oral treatment of rats with vanadate also increased rat blood pressure (Steffan *et al.*, 1984). However, whether this is due to the direct effects of vanadate on smooth muscle remains to be determined. Vanadate infusions in dog or cat caused arterial hypertension, increased peripheral resistance and caused a marked reduction or coronary and renal blood flow. Large arteries (femoral and carotid) were not constricted (Borchard *et al.*, 1981; Larsen and Thomsen, 1980; Hom *et al.*, 1982).

III. The Use of Vanadium to Study Biological Systems

The large amount of data which has accumulated concerning the effects of vanadate in *in vitro* and *in vivo* systems has led to the use of this metal as a tool in the study of biological structures and processes. The major uses of vanadium are described below, as well as some problems which can arise from extrapolation of *in vitro* results to *in vivo* systems.

A. SYSTEMS IN WHICH VANADIUM HAS BEEN USED AS A TOOL

1. Protein Structure. The paramagnetic vanadium compounds have been used as a tool in studying protein structure by EPR for many years (Chasteen, 1981). Changes in protein conformation can be monitored by the changes in the vanadyl

EPR spectra of vanadyl-protein complexes. Recently other forms of spectroscopy such as ENDOR and ESEEM (Chapter XI) and ^{51}V-NMR (Chapter X) have also been used to study structure. These studies are usually done on purified proteins.

2. Characterization of ATPases. Vanadate sensitivity/resistance has been used to characterize the multiple ATPase activities of the cell. Vanadate is a potent inhibitor of the Na,K ATPase of the plasma membrane, while it is not an inhibitor of the FlFo ATPAse of the mitochondrial membrane. The Na, K ATPase and the Ca^{++} ATPase of sarcoplasmic reticulum are the best studied of those enzymes which adopt two major conformation changes, E1-E2, in their catalytic cycle. Vanadate-inhibition has been used to classify an enzyme as an E1-E2 type. However, it must be noted that some of these enzymes, such as the fungal plasma membrane proton ATPase, have not been shown to be of this class using kinetic analysis. Since membrane cross-contamination is a problem in biological studies, the vanadate-sensitivity of ATPase activity has become a very useful tool and marker for plasma membrane activity which may be assayed in the presence of mitochondrial contamination. Further details of the inhibition of ATPases by vanadate are discussed in Chapter III.

The inhibitory affect of vanadate on ATPases has also been used to manipulate experimental conditions. For example, vanadate has been used to inhibit Ca^{++} uptake in secretion studies using the rat parotid gland. In this system vanadate is believed to act by inhibiting the Ca^{++} ATPase (Thevenod and Schulz, 1989).

3. The Study of Phosphorylation/Dephosphorylation Systems. Vanadate has been shown to interact with phosphorylation/dephosphorylation systems both *in vitro* and *in vivo*. Vanadate stimulation of cell proliferation is accompanied by increased levels of phosphorylated proteins. In general, the hormone-mimetic effects of vanadate on cells have all been associated with increased levels of phosphoproteins. *In vitro* analysis of specific enzymatic activities has not demonstrated a reproducible direct stimulation of tyrosine kinase activity, although increases in threonine and serine protein phosphorylation have been shown (Ennulat and Neff, 1985). Details of effects of vanadate on protein phosphorylation/dephosphorylation systems are discussed in Chapter IV.

B. POTENTIAL PROBLEMS USING OXOVANADIUM COMPOUNDS IN BIOLOGICAL
   SYSTEMS

The multiple interactions of oxovanadium compounds has made them useful tools in biological studies as probes, markers and treatments. However, the chemistry of oxovanadium compounds in solution is extremely complex, involving many pH and concentration dependent equilibria. In addition, the interconversion of oxovanadium compounds is greatly influenced by changes in oxidation-reduction conditions and vanadium compounds will bind to various components of biological systems. Vanadates can be useful tools to study biological systems as long as one is cognizant of their chemistry.

1. Interconversion of Oxovanadium Compounds during the Experiment. Oxovanadium compounds can interconvert during the course of an experiment. In experiments involving cells, vanadate(V) has been shown to be rapidly converted to vanadyl(IV). Glutathione and catechol are among the cellular constituents known to convert vanadate to vanadyl. In addition, cells contain various compartments with pH values ranging from 5.0 to 7.5 in which vanadium compounds may be concentrated. In this pH range vanadate can be converted to decavanadate. Vanadyl readily complexes with media or buffer constituents, making it more resistant to oxidation, but it can still be oxidized to vanadate albeit more slowly. In cell extract work vanadate(V) can be easily reduced to vanadyl(IV), or vanadyl(IV) oxidized to vanadate(V) depending on the buffer composition and state of aeration. In addition, there may be a distribution of various types of oxovanadium compounds present throughout cellular compartments.

2. Binding of Oxovanadium Compounds to Other Cellular Components. Vanadyl and vanadate both will bind to proteins, although vanadyl binding has higher association constants. The anionic vanadate compounds can bind to anionic components of biological buffers such as EDTA (Kustin and Toppen, 1973), citrate, and succinate (Willsky, unpublished observations). Therefore, the free concentration of oxovanadates in cells or extracts will vary as a function of other components in the system.

3. Difficulty in Assigning a Specific Oxidation State to the Oxovanadium Compound Active in Biological Systems. Magnetic resonance spectrometers, can unambigously determine the oxidation state of vanadium in an *in vitro* experiment. Care must be taken in extrapolating results obtained *in vitro*, under conditions in which the active oxovanadium compound has been identified, to those obtained in cells where prevailing conditions may alter the form and amount of free oxovanadium compounds.

C. GUIDELINES TO THE USE OF OXOVANADIUM COMPOUNDS IN CELL BIOLOGY

1. Consider the Chemically Predicted Form of Oxovanadium When Planning Experiments. At physiological pH = 7.5 vanadate exists in the metavanadate form. Purchased orthovanadate made up in water will be very alkaline (pH 12) while the commonly used vanadyl sulfate is very acidic (pH 2.0). When diluted into biological buffers vanadate will immediately convert to a mixture of monomeric and oligomeric vanadates depending on the concentration (cf. Table I). Care must be taken that buffers have adequate strength to counteract the alkalinity or basicity of the added oxovanadate. Vanadate solutions exist in equilibrium at neutral pH as mono-, di- and tetrameric species. Orthovanadate solutions (100 mM) in water are stable over long periods of time in the cold. We have found using ^{51}V-NMR that vanadate solutions prepared by diluting orthovanadate solutions have less impurities than equivalent metavanadate solutions made by using purchased metavanadate.

2. Do not store solutions containing vanadium and other components. Since oxovanadium compounds will establish equilibria dependent upon pH, vanadium concentration and concentration of potential ligands, they should be stored as distilled water solutions. In order to insure that the same form of vanadium is used in each experiment, directly dilute the stock vanadate water solution into the appropriate media or buffer at the start of each experiment.

3. Monitor the color of vanadate solutions. Orthovanadate should be a clear solution. Conversion to other forms of oxovanadium species can be observed by color changes. Reduction to vanadyl will cause a blue color, while conversion to the polymeric decavanadate will cause a yellow-orange color.

4. Consider the multiple roles of vanadium when interpreting data. Since various forms of vanadium either inhibit or stimulate various physiological processes, care must be taken in interpretation of results. In general, vanadate is the more inhibitory form of the metal, while vanadyl is the more stimulatory form. However, exceptions to this generalization do exist. Vanadate added to whole cell systems will be efficiently converted to vanadyl. Experiments studying phosphorylation/dephosphorylation are a good examples where vanadium chemistry, if properly applied, can be very useful. The formation of phosphorylated proteins can result from increased production of phosphoprotein or decreased breakdown of phosphoprotein. Vanadate, *in vitro* is a very potent inhibitor of protein phosphatases, while vanadyl *in vivo* appears to be a stimulator of protein phosphorylation. It is not clear if sufficient oxovanadium species, which inhibit phosphatase reactions, are present in living cells. The differential effects of interconvertable oxovanadium compounds on such systems should be carefully considered.

IV. Summary: The Role of Vanadium in the Study and Regulation of Biological Systems

The role of oxovanadium compounds in regulation and alteration of metabolism provides an opportunity to understand how inorganic chemistry processes can be involved in cellular regulation. Oxovanadium ions may be unique among the inorganic chemicals of biological interest in that they currently appear to have stimulatory pharmacological effects in of themselves and not in conjunction with a drug. In contrast, for example, the presence of metals is required for the action of bleomycin. The interactions of oxovanadium compounds with biological processes provides an interesting arena in which to study regulation processes utilizing both biological and chemical approaches. A generalized diagram of how oxovanadium compounds could interact with cellular components is shown in Figure 3. In extracellular fluids vanadate and vanadyl can be interconverted by oxidation and reduction. Both compounds can bind small metabolites and proteins, although vanadyl is more likely to bind to proteins. Either oxovanadium form could interact with extracellular domains of membrane proteins [illustrated

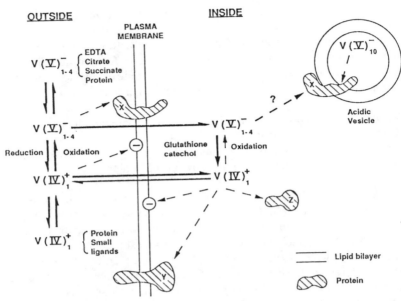

Fig. 3. Model for potential interactions of oxyvanadium compounds in cellular metabolism. Solid lines represent pathways which are known to exist. Dotted lines represent hypothetical interactions.

for V(V)] and affect intracellular metabolism. The cationic V(IV) is more likely to bind to negatively charged membrane phospholipid head groups than the anionic V(V). Vanadate and vanadyl have both been shown to enter the cell. The Km for vanadyl entry is lower, indicating that the affinity of vanadyl for its transporter is greater than that of vanadate(s) for its specific transporter under physiological conditions. Under phosphate starvation conditions, high affinity phosphate transport systems, which also recognize vanadates as substrates, are induced (Bowman and Slayer, 1979). Vanadate has been shown to enter via specific phosphate transport systems, or general anion transport systems, depending on the physiological conditions, while vanadyl is presumed to enter via monovalent cation transport systems. Once inside the cell vanadate is likely to be reduced to vanadyl by glutathione, catechol and other cellular components. It is possible, but unlikely, that vanadyl can be oxidized to vanadate inside the cell. Evidence of vanadyl efflux, but not vanadate, from the cell has been obtained. Inside the cell vanadyl can also interact with membrane lipids and both vanadate and vanadyl can interact with transmembrane or cytosolic proteins illustrated for V(IV). It is also possible that vanadate may be concentrated inside acidic vesicles causing the formation of decavanadate. The decavanadate could then interact with proteins inside the vesicle and cytosolic proteins after vesicle lysis.

Our knowledge of the role of oxovanadium compounds in normal metabolic processes remains incomplete. Oxovanadium compounds have certainly contributed to our knowledge of protein structure. Recent interest has focused on the effect of added oxovanadium compounds on metabolic processes. The inhibition of phosphohyrolases having a stable phosphoprotein complex in their catalytic cycle, the interaction of oxovanadium compounds with phosphorylation/dephosphorylation systems and their insulin-like effects have all contributed to the widespread use of vanadium in biological systems. Work is currently in progress in many laboratories exploring the affects of oxovanadium compounds on 2nd messenger systems, cellular calcium transport and mobilization, and the maintenance of intercellular pH. Continued work in these areas will eventually allow us to explain the physiological effects of oxovanadium compounds at the molecular level.

References

Akera, T., Temma, K. and Takeda, K.: 1983, *Fed. Proc* **42**, 2984–2988.

Al Bayati, M.A., Culbertson, M.R., Rosenblatt, L.S. and Hansen L.D.: 1982, *J. of Tox. and Environ. Health* **10**, 673–687.

Barrabin, H, Garrahan, J. and Rega, A.F.: 1980, *Biochim. Biophys. Acta* **600**, 796.

Benabe, J.E., Cruz-Soto, M.A. and Martinez-Maldonado, M.: 1984, *Am. J. Phys.* **246**, F317–322.

Bertrand, R.: 1950, *Bull. Amer. Mus. Nat. Hist.* **94**, 403

Blumenthal, N.C. and Cosma, V.: 1989, *J. Biomed. Mater. Res. App. Biomaterials* **23**, Al: 13–22.

Bond, G. and Hudgins, P.M.: 1980, *Biochim. Biophys. Acta* **600**, 796.

Borchard, U., Greeff, K., Hafner, D., Noack, E., Rojsathaporn, N.: 1981, *J. of Cardiovasc. Pharm.* **3**, 510–521.

Borchard, U., Fox, A.A., Greeff, K. and Schlieper, P.: 1979, *Nature (London)* **279**, 339–341.

Borst-Pauwels, G.W.F.H. and Peters, P.H.J.: 1981, *Biochim. Biophys. Acta* **642**, 173–184.

Bowman, B.J. and Slayman, C.W.: 1979, *J. Biol. Chem.* **254**, 2928.

Boyd, D.B. and Kustin, K.: 1984, *Adv. in Inorgan. Biochem.* **6**, 312–365.

Briskin, D.P., Thornley, W.R. and Poole, R. J.: 1985, *Archiv. Biochem. Biophys.* **236**, 228–237

Bruch, M., Dietrich, A. and Netter, K.J.: 1987, *Naunyn-Schmiedeberg's Arch. Pharmacol.* **336**, 111–116.

Campbell, C.A., Peet, M. and Ward, N. I.: 1988, *Biol. Psychiatry* **24**, 775–781.

Cantley, L.C. and Aisen, P.: 1979, *J. Biol. Chem* **254**, 1781–1784.

Cantley, Jr., L.C., Cantley, L.G. and Josephson, L.: 1978, *J. Biol. Chem.* **253**, 7361.

Carpenter, G.: 1980, *Biochem. Biophys. Res. Commun.* **102**, 1115–1121.

Carreras, J., Bartrons, R. and Grisolla, S: 1980, *Biochem. Biophys. Res. Commun.* **96**, 1267–1273.

Catalan, R.E., Martinez, A.M., Aragones, M.D., Robles, A. and Miguel, B.G.: 1987, *Life Sciences.* **40**, 799–808.

Catalan, R.E., Martinez, A.M., Aragones, M.D., Godoy, J.E. Robles, A. and Miguel, B.G.: 1982, *Biochemical Med.* **28**, 353–357.

Chasteen, N.D.:1981 in Biological Magnetic Resonance, Vol. 3, eds. Berliner, C. and Reuben, J., p. 53, Plenum Press, NY.

Chasteen, N.D.: 1983, *Structure and Bonding* **53**, 104–138.

Choate, G. and Mansour, T.E.: 1979, *J. Biol. Chem.* **254**, 11457–11462.

Churchhill, P.C. and M.C.: 1980, *J. Pharm.Exp. Ther.* **213**, 144–149.

Clausen, T. Andersen, T.C., Sturup-Johansen, M. and Petkova, O.: 1981, *Biochim. Biophys. Acta* **646** 261–267.

Collett, M. and Belzer, S.K.: 1987, *J. Virology* **61**, 1593–1601.

Cotton, F.A. and Wilkenson, G., Advanced Inorganic Chemistry, 4th edition, John Wiley and Sons, New York, 1980.

Coulombe, R.A., Jr., Briskin, D.B., Keller, R.J., Thornley, W.R. and Sharma, R. P.: 1987, *Arch. Biochem. Biophys.* **255**, 267–273.

Crans, D.C., Bunch, R.L. and Theisen, L.A.: 1989, *J. Am. Chem. Soc.* **111(19)**, 7597–7601.

Darr, D., and Fridovich, I.: 1984, *Arch. Biochem. Biophys*, **232**, 562–565.

Day, H., Middendorf, D., Lukert, B., Heinz, A. and Grantham, M.: 1980, *J. Lab. Clin. Med.* **96**, 382–395.

deBoer, E. and Wever, R.: 1988, *J. Biol. Chem.* **263**, 12326–12332.

Degani, H., Gochin, M., Karlish, S.J.D. and Shechter,: 1981, *Biochemistry*, **20**, 5795–5799.

DeMaster, E.G. and Mitchell, R.A.: 1973, *Biochemistry* **12**, 3616–3621.

DiPolo, R. and Beauge, L.: 1981, *Biochim. Biophys. Acta* **645**, 229.

DiPolo, R., Rojas, H.R. and Beauge, L.: 1979, *Nature* **281**, 228–229.

Domingo, J.L., LLobet, J.M., Tomas, J.M. and Crobella, J.: 1985, *J. Appl. Toxicol* **5**, 418–421.

Domingo, J.L., LLobet, J.M., Tomas, J.M. and Corbella, J.: 1986, *J. Appl. Toxicol.*, **6**, 337–341.

Dubyak, G.R. and Kleinzeller, A.: 1980, *J. Biol. Chem.*, **255**, 5306–5312.

Earp, H.S., Rubin, R.A., Austin, K.S. and Dy, R.C.: 1983, *FEBS* **161**, 180–184.

Edwards, R.M. and Grantham, J.: 1983, *Am. J. Physiol.* **245**, F772–F777.

English, L., Macara, I.G. and Cantley, L.C.: 1983, *J. Cell Biol.* **97**, 1299–1302.

Ennulat, C. and Neff, N.: 1987, Abstracts of the Yeast Cell Biology Meeting, p. 177 (Cold Spring Harbor, NY)

Erdmann, E.: 1980, *Basic Res. Cardiol.* **75**, 411–412.

Erdmann, E., Krawietz, W., Phillipp, G., Hackbarth, I., Schmitz, W., Scholz, H. and Crane, F.L.: 1979, *Nature* **282**, 335–336.

Fuhrmann, G.F., Schwarz, W., Kersten, R. and Sdun, H.: 1985, *Biochim. Biophys. Acta* **820**, 223–234.

Gentleman, S. Reid, T.W. and Martensen, T.M.: 1987, *Exp. Eye Res.* **44**, 587–594.

Gherzi, R., Caratti, C. Andraghetti, G., Bertonini, S., Montmurro, A., Sesti, G. and Cordera, R.: 1988, *Biochem. Biophys. Res. Commun.* **152**, 1474–1480.

Gibbons, I.R. *et al.*: 1978, *Proc. Natl. Acad. Sci. USA* **75**, 2220–2224.

Goodno, C.C.: 1979, *Proc. Nat. Acad. Sci. USA* **76**, 2620.

Goodno, C.C. and Taylor, E.E.: 1982, *Proc. Nat. Acad. Sci. USA,* **79**, 21.

Hagenmeyer, A. *et al.*: 1980, *Basic Res. Cardiol.* **75**, 452.

Hajjar, J.J. Fucci, J.C., Rowe, W.A. and Tomicic, T.K.: 1987, *Proc. of Soc. for Exp. Biol. and Med.* **184**, 403–409.

Hansen, V., Aeseth, J. and Alexander, J.: 1982, *Arch. of Toxicol.* **50**, 195–202.

Hatch, M., Freel, R.W., Goldner, A.M. and Earnest, D.L.: 1983, *Biochim. Biophys. Acta* **732**, 699–704.

Heath, E. and Howarth, O.W.: 1981, *J. Chem. Soc. Dalton*, 503–506.

Heide, M., Legrum, W., Netter, K.J. and Furhmann, G.F.: 1983, *Toxicology* **26**, 63–71.

Higashine, H., Bolgden, J.D., Lavenhar, M.A., Bauman, J.W., Jr., Hirotsu, T. and Aviv, A.: 1983, *Am. J. Physiol.* **244**, F105–F111.

Hom, G.J., Chelly, J.E. and Jandhyala, B.S.: 1982, *Proc. Soc. Exp. Biol. Med.* **169**, 401–405.

Howarth, O. W. and Jarrold, M. 1984, *J. Chem. Soc. Dalton*, pp. 503–506.

Hudson, T.O.F. 1964, *Vanadium: Toxicology and Biological Significance*, Elsevier Publishing Co., Amsterdam.

Inciate, D.J., Steffen, R.P., Dobbins, D. E., Swindall, B.T., Johnston, J. and Haddy, F.J.: 1980, *Am. J. Physiolog.* **239**, H47–H56.

Johnson, R.A.: 1982, *Arch. Biochem. Biophys.* **218**, 68–76

Johnson, T.M., Meisler, M.H., Bennett, M.I. and Willsky, G.R.: 1990, *Diabetes* **39**, 757–759.

Jones, M.M. and Basinger, M.A.: 1983, *J. of Toxicol. and Environmental Health* **12**, 749–756.

Kanagasuntheran, P. and Theo, T.C.: 1982, *Biochem. J.* **208**, 789–794.

Kanaho, Y., Chang, P.P., Moss, J. and Vaughn, M.: 1980, *J. Biol. Chem.* **263**, 17584–17589.

Klarlund, J.K., Latini, S. and Forchhammer,: 1988, *Biochemica Biophysica Acta* **971**, 112–120.

Kobayashi, *et al.*, 1978: *Biochem. Biophys. Res. Commun.* **81**, 1313.

Kountz, P.D., McCain, R.W., El-Maghrabi, M.R. and Pilkis, S.J.: 1986, *Arch. Biochem. Biophys.* **251**, 104–113.

Krawietz, W., Downs, R.W. Jr., Spiegel, A.M. and Aurbach, G.D.: 1982, *Biochem. Pharmacol.* **31**, 843–848.

Krivanek, J.: 1988, *Neurochemical Research* **13**, 395–401.

Kustin, K.: 1990, Biochemistry of Vanadium, Biochemistry of the Element Series, Ed. E. Frieden, Plenum Press, N.Y., in preparation.

Kustin, K. and McLeod, G.C.: 1983, *Structure and Bonding* 53, 139–160.

Kustin, K. and Toppen, D.L.: 1973, *J. Am. Chem. Soc.* 95, 3564–3568.

Labonnette, D.: 1988, *J. Chem. Res (S)*, 92–93.

Larmuth, L.: 1877, *J. of Anat. (London)* 11, 251–254.

Larsen, J.A. and Thomsen, O.O.: 1980, *Basic Res. Cardiol.* 75, 428–432.

Lau, K.H.W., Tanimoto, H.J. and Baylink, D.J.: 1988, *Endocrinology* 123, 2858–2867.

Lee, K.P. and Gillies, P.J.: 1986, *Environmental Research* 40, 115–135.

Legrum, W.: 1986, *Toxicology* 42, 281–289.

Lindquist, R.N., Lynn, J.L. Jr. and Lienhad, G.E.: 1973, *J. Am. Chem. Soc.* 95, 8762–8768.

Liochev, S. and Fridovich, I.: 1985, *J. of Free Rad. in Biol. and Med.* 1, 287–292.

Liochev, S. and Fridovich, I.: 1987, *Arch. of Biochem. and Biophys.* 255, 274–278.

Lopez, V., Stevens, T. and Lindquist, R.N.: 1976, *Arch. Biochem. Biophys.* 175, 31.

Lyonett, M.: 1899, *Presse Medicale* 1, 191–192.

Macara, I., Kustin, K. and Cantley, L.C.: 1980, *Biochim. Biophys. Acta* 629, 95–106.

Marini, M., Zunica, G., Bagnara, G.P. and Frencheschi, C.: 1987, *Biochem. Biophys. Res. Commun.* 142, 836–842.

Martin, D.M. and Chasteen, N.D.: 1988, *Meth. in Enz.* 158, 402–421.

Menon, A.S., Rau, M., Ramasarma, T. and Crane, F.L.: 1980, *FEBS Letters* 114, 139–141.

Minasi, L., Chang, A. and Willsky, G.R.: 1990, *J. Biol. Chem.* (in press).

Molnar, E., Kiss, Z., Dux, L. and Guba, F., 1988, *Acta Biochim. Biophys. Hung* 23(1), 63–74.

Mooney, R.A., Bordwell, K.L., Luhowskj, J.S. and Casnellie, J.E.: 1989, *Endocrinology* 124, 422–429.

Naylor, G.J. Corrigan, F.M., Smith, A.H.W., Conenlly, P. and Ward, N.I.: 1987, *Br. J. of Psychiatry*, 656–661.

Nechay, B.R.: 1984, *Ann. Rev. Toxicol.* 24, 501–524.

Nechay, B.R., Nanninga, L.B., Nechay, P.S.E., Post, R.L., Branthan, J.J., Macara, I.G., Kubena, L.F., Phillips, T.D. and Nielsen, F.H.: 1986, *Fed. Proc.* 45, 123–132.

Nordmann, J.J. and Zyzek, E.: 1982, *J. Physiol.* 325, 281–299.

North, P. and Post, R.L.: 1980, *J. Biol. Chem.* 259, 4971–4978.

Paris, A. and Poussegur, J.: 1987, *J. Biol. Chem.* 262, 1970–1976.

Patterson, B.W., Hansard II, S.L., Ammerman, C.B., Henry, P.R., Zech, L.A. and Fisher, W.R.: 1986, *Am. J. Physiol* 251, R325–332.

Pochal, M., Acara, M., Lohr, J., McReynolds, J., Bennett, M.I. and Willsky, G.R.: 1989, *Diabetes* 38,S2, 133A

Priestly, 1877, *Philosophial transactions of the Royal Society (London)*, Part 2, 495–556.

Priyadarshini, U. and Tandon, S.G.: 1961, *Anal. Chem.* 33, 435–438.

Quist, E.E. and Hokin, L.E.: 1978, *Biochim. Biophys. Acta* 511, 202.

Ramanadham, M. and Kern, M.: 1983, *Molec. Cell Biochemistry* 51, 67–71.

Ramanadham, S., Mongold, J.J., Brownsey, R.W., Cros, G.H. and McNeill, J.H.: 1989, *Am. J. Physiol.* 257, H904–H911.

Ramasarma, T. and Crane, F.L.: 1981, *Curr. Topics in Cell. Reg.* 20, 247–299.

Ramasarma, T., MacKeller, W.C. and Crane, F.L.: 1981, *Biochim. Biophys. Acta* 646, 88–98.

Reisman, H.: 1985, *Cell* 40, 1001–1009.

Ryder, J.W. and Gordon, J.A.: 1987, *Molec. Cell Biology* 7, 1139–1147.

Sato, B., Miyashita, Y., Maeda, U., Noma, K., Kismimoto, S. and Masumoto, K.: 1987, *Endocrinology* 120, 1112–1120.

Schmitz, W., Scholz, H., Erdmann, E., Krawietz, W. and Werdan, K.: 1982, *Biochemical Pharmacology* 31, 3853–3860.

Schwartz, A. *et al.*: 1980, *Basic Res. Cardiol.* 75, 444 and references therein.

Seargeant, L.E. and Stinson, R.A.: 1979, *Biochem. J.* 181, 247.

Shechter, Y. and Karlish, S.J.D.: 1980, *Nature* 284, 556–558.

Shimizu, T., 1981: *Biochemistry* 20, 4347–4354.

Smith, B.E., Eady, R.R., Lowe, D.J. and Gormal, C.: 1988, *Biochem. J.* 250, 299–302.

Steffan, R.P., Pamnani, M.B., Cloush, D.L., Muldoon, S.M. and Haddy, F.J.: 1981, *Hypertension* **3**, 1173–1178.

Sundet, W.D., Wang, B.C., Hakumaki, M.O.K. and Goetz, K.L.: 1984, *Proc. Soc. Exp. Bio. Med.* **175**, 185–190.

Swarup, G., Cohen, S. and Garbers, D.L.: 1983, *Biochem. Biophys. Res. Commun.* **107**, 1104–1109.

Tanaka, K., Waxman, L. and Goldberg, G.: 1984, *J. Biol. Chem.* **259**, 2803–2809.

Takeda, K., Akera, T., Yamamoto, S. and Shieh, I-S.: 1980, *Naunyn-Schmiedeberg's Arch. Pharmacol.* **314**, 161–170.

Tamura, S., Brown, T., Dubler, R.E. and Larner, J.: 1983, *Biochem. Biophys. Res. Commun.* **113**, 80–85.

Tamura, T., Simon, E., Niemann, H., Snoek, G. and Bauer, H.: 1986, *Molec. Cell Biology* **6**, 4745–4748.

Thevenod, F. and Schulz, I.: 1988, *Am. J. Phys.* **255**, G429–G440.

Thompson, H.J., Chasteen, N.D. and Meeker, L.D.: 1984, *Carcinogenesis* **5**, 849–851.

Tojo, A., Kasuga, M., Urabe, A. and Takaku, F.: 1987, *Experimental Cell Research* **171**, 16–23.

Tracy, A.S. and Gresser, M.J.: 1986, *Proc. Natl. Acad. Sci.* **83**, 609–613.

Tracey, A.S., Gresser, M.J. and Liu, S.: 1988, *J. Am. Chem. Soc.* **110**, 5869–5874.

Tucker, P.W., Hazon, Jr., E.E. and Cotton, F.A.: 1979, *Mol. Cell Biochem.* **23**, 67.

Van Etten, R.L., Waymack, P.P. and Rehkop, D.M.: 1974, *J. Am. Chem. Soc.* **96**, 6782.

Varecka, L. and Carafoli, E.: 1982, *J. Biol. Chem.* **257**, 7414–7421.

Vijaya, S., Crane, F.L. and Ramasarman, T.: 1984, *Molec. and Cell. Biochem.* **62**, 175–185.

Vyskocil, F., Teisinger, J. and Dlouha, H.: 1981, *Biochem. Biophys. Res. Commun.* **100**, 982.

Willsky, G.R., White, D.A. and McCabe, B.C.: 1984, *J. Biol. Chem.* **259**, 13273–13281.

Willsky, G.R., Offermann, P.U., Jr., Plotnick, E.K., Dosch, S.F. and Leung, J.O.: 1985, *J. Bacteriol.* **164**, 611–617.

II. The Coordination and Redox Chemistry of Vanadium in Aqueous Solution

ALISON BUTLER

Department of Chemistry, University of California, Santa Barbara, CA 93106, U.S.A.

With the recent discoveries of the first two naturally occurring vanadium-containing enzymes, the bromoperoxidase and nitrogenase, the coordination chemistry and reactivity of vanadium is receiving renewed attention. One of the most intriguing features of vanadium is the number of stable oxidation states in which vanadium is known and the relative ease of interconversion between many of these states. Vanadium complexes in the oxidation states, -3, -1, 0, $1+$, $2+$, $3+$, $4+$ and $5+$ have been well characterized, although, only the $2+$ through $5+$ states are well known in aqueous solution and the $3+$ through $5+$ are the oxidation states in which vanadium is found in biological systems.

This chapter will focus primarily on examples of vanadium(II, III, IV, V) complexes in aqueous solution that undergo electron transfer reactions or that catalyze redox reactions, such as the oxidation of halides by hydrogen peroxide, which is catalyzed by vanadium bromoperoxidase (see Chapter V) or the reduction of dinitrogen, which is catalyzed by vanadium nitrogenase (see Chapter VI). This article is not intended to be a comprehensive summary of the coordination and redox chemistry of vanadium. For a recent summary of vanadium coordination chemistry, the reader is referred to Boas and Pessoa's (1987) article in Comprehensive Coordination Chemistry, Volume 3. Other reviews that cover aspects of the electron transfer and coordination chemistry of vanadium or that examine the biological role of vanadium include Saito and Sasaki, 1988; Holm, 1987; Boyd and Kustin, 1984; Nechay, 1984; Chasteen, 1983; Kustin *et al.*, 1983; Kustin and Macara, 1982; Chasteen, 1981; and Ramasarma and Crane, 1981.

1. Coordination Geometries and Reduction Potentials of Vanadium Complexes in Aqueous Solution

The reduction potentials and the predominant oxidation states of vanadium vs pH are summarized in Table I and Figure 1. As is apparent from Table I and Figure 1, the oxyanion, vanadate(V), is the most prevalent species present in aqueous solution at neutral pH. Vanadate readily undergoes polymerization reactions depending on its concentration, the pH and the nature of other ions present in solution. For example, millimolar vanadate solutions at pH 7 (Hepes buffer) are

N. Dennis Chasteen (ed.), Vanadium in Biological Systems, 25–49.
© 1990 *Kluwer Acadmic Publishers. Printed in the Netherlands.*

Table I. Standard reduction potentials of vanadium.

$E°$	Strongly acidic solutions	Neutral to basic solutions (pH 6–12)
V^V–V^{IV}	1.000	0.991
V^{IV}–V^{III}	0.337	0.542
V^{III}–V^{II}	−0.255	−0.486
V^{III}–V^{II}	−1.13	−0.820

Data obtained from Isreal and Meites (1985).

predominantly composed of $H_2VO_4^-/HVO_4^{2-}$, $H_2V_2O_7^{2-}$, and $V_4O_{12}^{4-}$ (Tracey and Gresser, 1988):

$$4\,H_2VO_4^- \longrightarrow V_4O_{12}^{4-} + 4H_2O \qquad \sim 3 \times 10^8\ M^{-3}$$
$$2\,H_2VO_4^- \longrightarrow H_2V_2O_7^{2-} + H_2O \qquad \sim 3 \times 10^2\ M^{-1}$$
$$2\,H_2V_2O_7^{2-} \longrightarrow V_4O_{12}^{4-} + 2H_2O \qquad \sim 3 \times 10^3\ M^{-1}$$

Higher order oligomers, including decavanadate, $H_2V_{10}O_{28}^{4-}$, are formed at a higher vanadate concentration or lower pH. Recent work by Crans and Shin (1988) and Tracey and Gresser (1988) has demonstrated that the choice of buffer

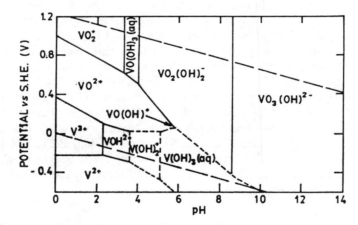

Fig. 1. Reduction potential E (referenced to the standard hydrogen electrode versus pH for various species of vanadium. Boundary lines correspond to E-pH values where the species in adjacent regions are present in equal concentrations. The short dashed lines indicate uncertainty in the location of the boundary. The *upper* and *lower long dashed lines* correspond to the upper and lower limits of stability of water. Standard reduction potentials are given by the intersections of "horizontal" lines with the abscissa pH = 0. The half reactions are $O_2 + 4H^+ + 4e = 2H_2O$, $E° = 1.23V$; $VO_2^+ + 2H^+ + e = VO^{2+} + H_2O$, $E° = 1.0V$; $VO^2 + e = V^3 + H_2O$, $E° = 0,36V$; $2H^+ + 2e = H_2$, $E° = 0.0V$; and $V^{3+} + 2e = V^{2+}$, $E° = 0,25V$. V^{2+} is therefore a strong reductant. Air oxidation of VO^{2+} presumably proceeds by the reaction $4VO^{2+} + O_2 + 2H_2O = 4VO_2^+ + 4H^+$, $E° = 0.23V$ which is favored at higher pH. Not all known species are represented on this diagram.

ion critically affects the nature of the vanadate species present in solution. Hepes buffer (4-(2-hydroxyethyl)-1-piperazineethanesulfonic acid) appears to be one of the least coordinating buffer salts, whereas other multidendate buffer ions (e.g., Tris, Bicine, Bis-tris and *many* others) complex vanadate with much greater binding constants (Crans and Shin, 1988; Tracey and Gresser, 1988; Tracey *et al.*, 1988).

While vanadate is the most stable oxidation state in aqueous solution at neutral pH, lower oxidation state complexes are certainly found. The nature of the coordinating ligand plays a particularly important role in stabilizing the lower oxidation state complexes against hydrolysis and oxidation.

Vanadium complexes adopt a wide variety of coordination geometries as briefly summarized by the examples in Table II.

The most common coordination mode for vanadium(II) and vanadium(III) complexes is octahedral. The molecular geometries adopted by vanadium(IV) and

Table II. Oxidation states and stereochemistries of vanadium complexes.

Oxidation state	Coordination number	Stereochemistry	Examples
− 3	6		$(Et_4)_2[(Ph_3Sn)V(CO)_5]$
	7	Monocapped octahedral	$(Et_4N)[Ph_3Sn)_2V(CO)_5]$
− 1	6	Octahedral	$[V(bipy)_3]^-$
0	6	Octahedral	$[V(bipy)_3]$
1	6	Octahedral	$[V(bipy)_3]^+$
Low but not defined	6	Trigonal prismatic	$[V(S_2C_2Ph_2)_3]$
2	6	Octahedral	$[V(CN)_6]^{4-}$, $[V(H_2O)_6]^{2+}$
3	3	Planar	$[V\{N(SiMe_3)_2\}_3]$
	4	Tetrahedral	$[VCl_4]^-$
	5	Trigonal bipyramidal	trans-$[VCl_3(NMe_3)_2]$
	6	Octahedral	$[V(H_2O_6]^{3+}$, $[V(C_2O_4)_3]^{3-}$
	7	Pentagonal bipyramidal	$K_4[V(CN)_7]\cdot 2H_2O$
4	4	Tetrahedral	$[VCl_4]$
	5	Trigonal bipyramidal	$[VOCl_2(NMe_3)_2]$
		Square pyramidal	$[VO(acac)_2]$, $[VS(acac)_2]$
	6	Octahedral	$[VO(acac)_2(dioxane)]$, $[V(catecholate)_3]^{2-}$, $[V(acac)_2Cl_2]$
	8	Dodecahedral	$[V(MeSC_3)_4]$, $[VCl_4(diars)_2]$
5	4	Tetrahedral	$[VOCl_3]$, $[V(Me_3SiO)_3(adamantylimido)]$
	5	Trigonal bipyramidal	VF_5 (g), VCl_5 (g)
		Square pyramidal	$[VOF_4]^-$
	6	Octahedral	$[VO(OMe)_3]$, $[VO_2(ox)_2]^{3-}$, $[VOCl_3(benzonitrile)_2]$
	7	Pentagonal pipyramidal	$[VO(NO_3)(MeCN)]$, $[V(dipic)(H_2NO)(NO)(H_2O)]^-$
	8	Dodecahedral	$[V(O_2)_4]^{3-}$

Reproduced with permission from Boas and Pessoa (1981).

vanadium(V) complexes, however, are far more flexible. Several interesting or novel coordination modes of vanadium(IV) and vanadium(V) warrant greater discussion. These include ligand stabilization of bare V^{4+} and even bare V^{5+} ions, the organic functional group analogy to several vanadium(IV and V) coordination complexes, vanadium(V) peroxide complexes in light of V-bromoperoxidase and thiobridged vanadium-iron complexes in light of V-nitrogenase.

The coordination chemistry of vanadium(IV) is usually considered to be dominated by the vanadyl, $V = O^{2+}$, moiety and vanadium(V) by the VO^{3+} and cis-VO_2^+ moieties. To be sure, a majority of vanadium(IV) and vanadium(V) complexes contain these functionalities, although, Floriani and coworkers were the first to demonstrate that the VO^{2+} moiety is indeed more reactive than once generally believed. As presented in more detail below, the oxide ligand of several V^{IV}-*bis*bidentate chelate complexes (VO(chel)$_2$; chel = acetylacetonate, 8-hydroxyquinolinate, N-n-butyl{salicylideniminate} and N,N'- ethylenebis{salicylidene}) can be displaced by reaction with more oxophilic reagents, i.e., $SOCl_2$, PCl_5 (Pasquali *et al.*, 1979) or in strongly acidic media (Bonadies *et al.*, 1987), forming stable complexes of V(chel)$_2$Cl$_2$ in dry organic solvents. Other examples of bare vanadium(IV) complexes, i.e., those lacking the coordinated oxide ligand, which were once considered to be rare, are increasing, some of which are even stable in aqueous solution. The tris 1,2 dicyanoethylene-1,2-dithiolate complex of V(IV), which was the first example of a bare V^{4+} ion, possesses a trigonal prismatic coordination geometry (Stiefel *et al.*, 1967). In addition to this and other V^{IV}(dithiolene)$_3^{2-}$ complexes (Matsubayshi *et al.*, 1988; Welch *et al.*, 1988), V^{IV}(catecholate)$_3^{2-}$ complexes (Cooper *et al.*, 1982), V^{IV}(acetylacetonate)$_3^+$ complexes (Hambley *et al.*, 1987; Diamantis *et al.*, 1980) and mixed ligand V^{IV}(catecholate)(acetylacetonate)$_2$ complexes (Hawkins and Kabanos, 1989) have been well characterized, maintaining either trigonal prismatic or octahedral coordination (Figure 2). Perhaps the most prominent example of a bare V(IV) complex is the hexa-amine vanadium(IV) cage complex [V^{IV}{di(amH)sar-2H}]-$(S_2O_6)_2$, (di(amH)sar-2H = 1,8-di-aminosarcophagine = 3,6,10,13,16,19-hexaaza-bicyclo [6.6.6]eicosane) prepared by Sargeson and coworkers which is a saturated hexa-amine complex of V(IV), stable in aqueous solution (see Figure 2; Comba *et al.*, 1985). It has a nearly trigonal prismatic structure, coordinated by two deprotonated amide ligands and four protonated amine ligands (Comba *et al.*, 1985). Thus clearly the bare V^{4+} ion is a viable entity even in aqueous solution. Even the bare V^{5+} center has been stabilized by tris catecholate coordination (Cooper *et al.*, 1982).

ORGANIC FUNCTIONAL GROUP ANALOGY

Floriani *et al.* have drawn analogies between the reactivity of oxo-vanadium (IV and V) complexes and organic functional groups (Giacomelli *et al.*, 1982). The complex L_2V^{IV}-μ-O-$V^{IV}L_2$ (L = 8-hydroxyquinolinate, Figure 3I) undergoes esterification to form $L_2V(O)(OR)$ (see Figure 3V), base hydrolysis to form L_2VO_2 (Figure 3II), and acylation to form $L2V(O)(OC(CH_3)O)$ (Figure 3IV). The cis-

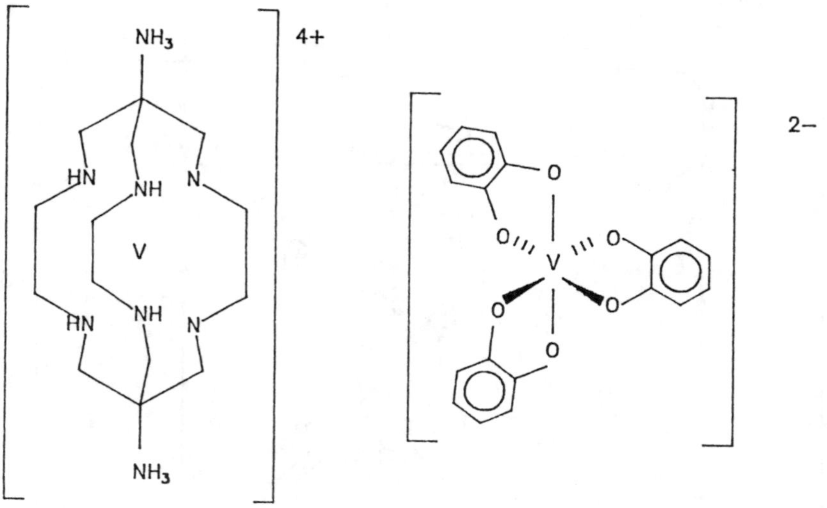

Fig. 2. Structures of $[V^{IV}\{di(amH)sar-2H\}](S_2O_6)_2$, (di(amH)sar-2H = 1,8-di-aminosarcophagine = 3,6,10,13,16,19-hexaaza-bicyclo [6.6.6]eicosane; Comba et al., 1985), a trigonal prismatic coordination and $V^{IV}(catecholate)_3^{2-}$ complexes (Cooper et al., 1982), showing octahedral coordination.

dioxo $[VO_2]^-$ moiety of Figure 3II can also be protonated (Figure 3III) forming a carboxylic acid-like complex (Giacomelli et al., 1982). A more recent functional group analogy is to the crown ethers. Pecoraro (1989) has reported the crystal structure of $[VO(salicylhydroxamate)(CH_3OH)]_3$ (Figure 4). The cluster forms a $[-V-N-O-]_3$ core which originates from coordination of the hydroxamate oxygens to one VO^{3+} unit and coordination of the deprotonated hydroxamate nitrogen and phenolate oxygen to another VO^{3+} unit. In analogy to the cation binding properties of conventional crown ethers, Pecoraro and coworker are investigating the cation transport processes of $[VO(salicylhydroxamate)(CH_3OH)]_3$ (Lah et al., 1990). The resemblance of certain vanadium(V) peroxide complexes to alkyl peroxides is discussed below.

VANADIUM(V) PEROXIDE COMPLEXES

Vanadium bromoperoxidase is a vanadium(V) containing enzyme that catalyzes the oxidation of bromide or iodide by hydrogen peroxide (See Chapter VI). The reaction is thought to proceed by peroxide coordination to vanadium(V). Hydrogen peroxide readily coordinates to vanadium(V) complexes as has been demonstrated by the kinetic and mechanistic work of several groups, including Orhanovic and Wilkins (1967), Wieghardt (1978), Tanaka et al. (1980), etc. The majority of all vanadium(V) peroxide complexes adopt seven coordinate, pentagonal bipyramidal geometries as displayed by the structure of $VO(O_2)(Pic) \cdot 2H_2O$ (Pic = pyridine-2-carboxylate; Figure 5; Mimoun et al., 1983). The oxide ligand of VO is situated at the axial position. Other ligands which form seven coordinate

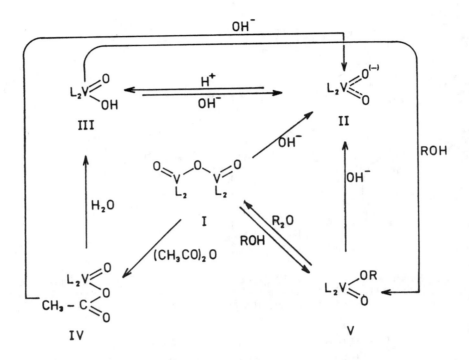

L = 8-quinolinato anion

Fig. 3. The functional group analogy of the oxovanadium(V) complex L_2V^{IV}-μ-O-$V^{IV}L_2$ (L = 8-hydroxyquinolinate). Reproduced with permission from Giacomelli *et al.* (1982).

oxovanadium-peroxide complexes include pyridine-3-carboxylic acid (nicotinic acid; Djordjevic *et al.*, 1988), citric acid (Djordjevic *et al.*, 1989), iminodiacetic acid (Djordjevic *et al.*, 1985), bipyridine (Sventivanyi and Stomberg, 1983), dipicolinic acid (Drew and Einstein, 1973), oxalate (Stomberg, 1986) etc. A notable exception is the structure of $NH_4[VO(O_2)_2(NH_3)]$ which has a distorted pentagonal pyramidal structure (Figure 5; Drew and Einstein, 1972). The peroxide complexes are remarkably stable in the dark.

THIOBRIDGED VANADIUM-IRON COMPLEXES

One of the protein components of the vanadium nitrogenase system, contains a vanadium-iron-sulfur extractable cofactor which partially resembles the molybdenum-iron-sulfur cofactor of the conventional nitrogenase system (see Chapter VII). Thus interest in thiobridged vanadium-iron clusters has soared. While these complexes are not stable in aqueous solution, recent synthetic developments are presented here because of their relevance to the vanadium nitrogenase cofactor. Kovacs and Holm (1986, 1987) have recently shown that the cubane-type cluster, $(Me_4N)VFe_3S_4Cl_3(DMF)_3 \cdot 2DMF$, is readily prepared by reaction of the linear,

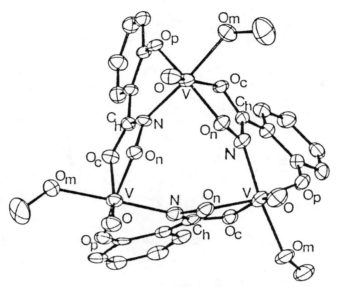

Fig. 4. An ORTEP diagram of [VO(salicylhydroxamate)(CH$_3$OH)]$_3$. Reproduced with permission from Lah *et al.* (1990). (Thermal ellipsoids at 30% probability. Selected mean bond lengths (Å) and angles for chemically equivalent bonds: V = O, 1.59; V-O$_p$, 1.85; V-O$_m$ 2.08; V-O$_c$ 2.12; V-O$_n$ 1.86; V-N, 2.02; O$_n$-N 1.37; C$_h$-N 1.33; C$_h$-O 1.26; V-V 4.66; O = V-O$_m$ 91°; O = V-O$_p$ 108°; O = V-O$_n$ 94°; O = V-N 98°; O = V-O$_c$ 165°; V-N-O$_n$ 120°; N-O$_n$-V 120°.)

trinuclear [VFe$_2$S$_4$Cl$_4$]$^{3-}$ cluster with FeCl$_2$ or by reaction of tetrathiovanadate and FeCl$_2$ in DMF (Figure 6). The (VFe$_3$S$_4$)$^{2+}$ core is isoelectronic with the (MoFe$_3$S$_4$)$^{3+}$ core and nearly isostructural, as well. The ^{57}Fe Mossbauer spectrum and the bond lengths (Fe-Cl, V-O$_{DMF}$) suggest the VFe$_3$S$_4$ cluster is best characterized as a delocalized distribution of oxidation states, e.g.,

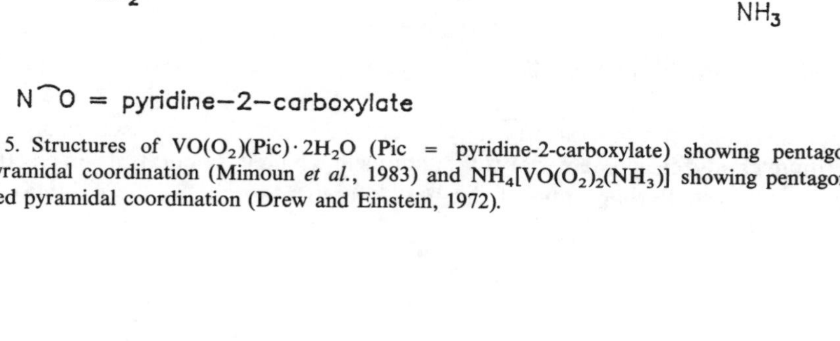

N⌒O = pyridine−2−carboxylate

Fig. 5. Structures of VO(O$_2$)(Pic)·2H$_2$O (Pic = pyridine-2-carboxylate) showing pentagonal bipyramidal coordination (Mimoun *et al.*, 1983) and NH$_4$[VO(O$_2$)$_2$(NH$_3$)] showing pentagonal-based pyramidal coordination (Drew and Einstein, 1972).

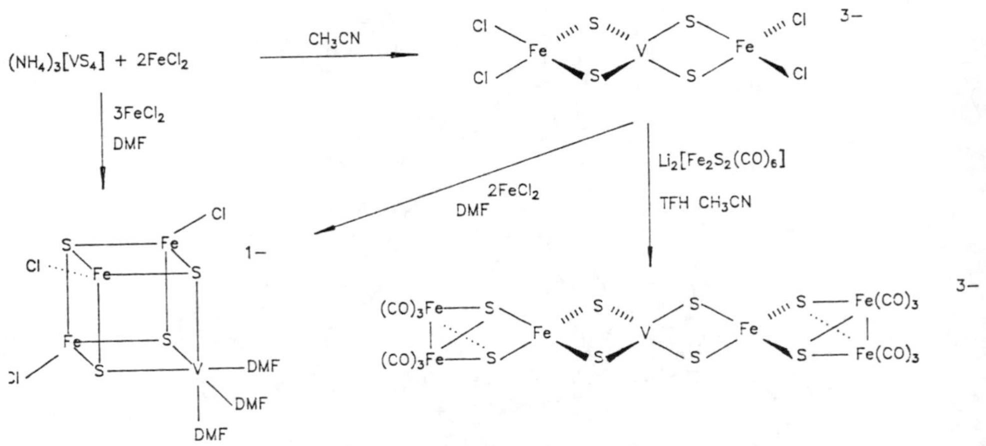

Fig. 6. Synthetic scheme for thiobridged vanadium-iron clusters (adapted from Kovacs and Holm, 1986, 1987).

V(2.5)/3Fe(2.5), as opposed to a localized charge distribution of V(IV)/3Fe(II). The magnetic susceptibility of $(Me_4N)[VFe_3S_4Cl_3(DMF)_3] \cdot 2DMF$ obeys near Curie behavior ($\mu = 3.87$), indicative of a S $= 3/2$ ground state. Some of these features are remarkably similar to the vanadium-iron-sulfur cluster extracted from vanadium-nitrogenase. The one-electron reduction potential of the $VFe_3S_4Cl_3(DMF)_3^-$ cluster is 0.12 V vs SCE. The terminal chloride and DMF ligands of $(Me_4N)VFe_3S_4Cl_3(DMF)_3 \cdot 2DMF$, have been replaced by thiolate, phosphene, phenoxide and bipyridine ligands (Kovacs and Holm, 1987a), suggesting that other novel cluster environments can be synthesized, possibly with exceptional reactivity. Holm has recently reported the insertion of $[VFe_3S_4]^{2+}$ into a trithiolate-containing cavitand ligand which coordinates at the iron sites, leaving the vanadium site open for ligand substitution (Ciurli and Holm, 1989). In addition to the nearly linear $[VFe_2S_3Cl_4]_3^-$ ($<$ Fe-V-Fe $= 172.9°$) complex and the VFe_3S_4 cubane cluster, other novel clusters are also known including the linear $[VFe_6S_8(CO)_{12}]^{3-}$ cluster (Kovacs and Holm, 1987). Ti_3VS_4 and the mineral sulvanite, Cu_3VS_4, are also examples of thiobridged vanadium-metal clusters present in extended lattice networks, although neither have been identified as discrete monomeric clusters. The mineral patronite $(VS_4)_n$ is a chain of pairs of vanadium(IV) ions bridged by side-on bound disulfide ligands (Halbert et al., 1986).

2. Redox Reactivity of Vanadium

Because vanadium can exist in many oxidation states, it is not surprising that vanadium catalyzes a wide variety of reactions in which it functions as an electron transfer catalyst. Perhaps one of the best known examples is the oxidation of sulfurous acid (SO_2) by vanadium, which is used in the industrial synthesis of sulfuric acid. The intent in this section is to describe other reactions that are catalyzed by vanadium, most of which function in an electron transfer capacity. The examples chosen have possible relevance to the role of vanadium in biological systems.

VANADIUM CATALYSIS OF SUPEROXIDE DISPROPORTIONATION

According to the reduction potentials given by Latimer (1952) or Isreal and Meites (1985), the vanadyl ion, VO^{2+}, has the potential to catalyze the disproportionation of superoxide, forming hydrogen peroxide and dioxygen in 1 M H^+.

$$2\,HO_2 \longrightarrow O_2 + H_2O_2 \qquad\qquad A$$

The reduction potentials are consistent with redox cycling between V(IV) and V(V) *or* between V(IV) and V(III), at pH 0 (Scheme 1). The overall driving force for the oxidation and reduction of VO^{2+} by HO_2 are nearly the same, i.e., 0.50 V and 0.49 V, respectively. The driving force for the reduction of V(V) by a second equivalent of HO_2 is 1.13 V and for the oxidation of V(III) is 1.17 V, confirming that both catalytic pathways are thermodynamically allowed for the vanadium-catalyzed disproportionation of superoxide at pH 0.

SCHEME 1

$$V^{IV} + HO_2 \longrightarrow V^{III} + O_2 \qquad or \qquad V^{IV} + HO_2 \longrightarrow V^{V} + H_2O_2$$
$$V^{III} + HO_2 \longrightarrow V^{IV} + H_2O_2 \qquad\qquad V^{V} + HO_2 \longrightarrow V^{IV} + O_2$$

Many first-row transition metal ions and their complexes catalyze the disproportionation of superoxide, including three metalloproteins, Cu, Zn-superoxide dismutase (SOD), Fe-SOD and Mn-SOD, so it is not surprising that the vanadyl ion is also a catalyst of superoxide disproportionation (Rush and Bielsky, 1985). It has even been speculated that a naturally occurring vanadium-containing superoxide dismutase might also be present in biological systems.

Rush and Bielski (1985) have shown that vanadium functions in the catalysis of superoxide disproportionation by cycling between vanadium(IV) and vanadium(V) (i.e., equations B and C). Curiously, the VO^{2+}/V(III) catalytic route was not observed.

$$VO^{2+} + HO_2 \longrightarrow VO_3^+ + H^+ \qquad\qquad B$$
$$VO_3^+ + HO_2 \longrightarrow VO^{2+} + O_2 + HO_2^- \qquad\qquad C$$

The reaction of VO^{2+} and HO_2 (Rxn B) occurs via the production of a transient intermediate, with the proposed composition, $VO^{2+} \cdot O_2H$. This transient undergoes a pH dependent deprotonation reaction, yielding VO_3^+. In 1 M acid, kb is 1.1×10^4 M^{-1} sec^{-1}. The VO_3^+ ion is the red monoperoxoV(O)$^+$ species, which is characterized by a λmax at 450 nm. Reaction of VO_3^+ and HO_2 (Rxn C) which is independent of pH between 0 and 1.65, produces dioxygen and hydrogen peroxide via the formation of the transient species $VO_3^+ \cdot O_2H$ (λmax 400 nm). The rate constant k_c can be calculated to be ca. $7 \times 10^4 \ M^{-1}sec^{-1}$, provided $K[O_2^-] \ll 1$, where K is the formation constant defined by $[VO_3^+ \cdot O_2H]/[VO_3^+][HO_2]$ (Rush and Bielski, 1985).

The reduction of $Cu^{II}Zn$-SOD and the oxidation of $Cu^I Zn$-SOD by superoxide occur at essentially diffusion controlled rates (i.e., $k \sim 10^9 \ M^{-1}sec^{-1}$). By comparison vanadium is not as efficient as Cu, Zn-SOD and other simple copper(II) complexes at catalyzing the disproportionation of superoxide. Nevertheless, V^{IV}/V^V system at pH 0 still provides a significant rate enhancement over the self-disproportionation of superoxide. In 1 M acid, the uncatalyzed disproportionation rate constant is calculated to be $7.5 \times 10^5 \ M^{-1}sec^{-1}$ but this is a *true* second order reaction that depends on the concentration of superoxide.

EDTA inhibits the reaction of VO^{2+} and HO_2. This result has been interpreted to mean that HO_2 does not reduce VO^{2+} in an outer sphere process (Rush and Bielski, 1985).

VANADATE CATALYZED OXIDATION OF NADH AND NADPH

Vanadate catalyses the oxidation of the reduced nicotinamide adenine dinucleotides, NADH and NADPH in biological systems. The phenomenon was first observed using cell and organelle membranes which exhibited a vanadate-dependent oxidation (Ramasarma *et al.*, 1981). Several groups have also reported this phenomenon (Patole *et al.*, 1986; Coulombe *et al.*, 1987; Liochev *et al.*, 1989; and references therein), however, considerable controversy has developed concerning the mechanism and the identity of the active catalytic component.

NAD(P)H   NAD(P)+

Fridovich *et al.* reasoned that the vanadate effect could be explained if membrane-associated NAD(P)H oxidases produced superoxide and if the superoxide reacted with an oxidant, such as vanadium(V), initiating a free radical chain oxidation of NAD(P)H. According to Fridovich, many molecules of NAD(P)H are oxidized per equivalent of superoxide (Darr and Fridovich, 1984,1985; Liochev and Fridovich, 1986). Liochev and Fridovich (1986; 1987) have reported that superoxide dismutase (SOD) inhibited the vanadate stimulated oxidation of NAD(P)H, while catalase had no effect. Fridovich's results were independent of the source of superoxide [e.g., the enzymatic xanthine oxidase system (XO; Darr and Fridovich, 1984, 1985; Liochev et al., 1989), chemical/photochemical (Liochev and Fridovich, 1986)], suggesting that superoxide is the species that reacts with vanadium. By contrast Khandke et al. (1986) argue that the V(V) dependent NADH oxidation in the presence xanthine/XO is an intrinsic property of the system. The vanadyl ion was also reported to be a catalyst in so much as it was autocatalytically oxidized to vanadate by dioxygen producing superoxide and by the production of hydroxyl radicals from reduction of hydrogen peroxide by the vanadyl ion (Liochev and Fridovich, 1987).

A mechanism proposed by Fridovich et al. that accounts for the vanadate stimulation of NAD(P)H oxidation by superoxide is summarized in Scheme 2 (Liochev, et al., 1989). Certain features of this mechanism will require further investigation, particularly hydride abstraction reactions by the proposed "vanadyl-dioxygen" species.

SCHEME 2

$$V^V + O_2^- \rightleftharpoons V^{IV}\text{-OO}$$
$$V^{IV}\text{-OO} + NAD(P)H \rightleftharpoons V^{IV}\text{-OOH} + NAD(P)^{\cdot}$$
$$V^{IV}\text{-OOH} + H^+ \rightleftharpoons V^V + H_2O_2$$
$$NAD(P)^{\cdot} + O_2 \rightleftharpoons NAD(P)^+ + O_2^-$$

$$NAD(P)H + O_2 H^+ \rightleftharpoons NAD(P)^+ + H_2O_2$$

Fridovich's results do not exclude the possibility that vanadate interacts directly with a membrane-oxidase, thereby effecting the oxidation of NAD(P)H. In recent studies, Reif et al. (1989) found that NADPH-cytochrome P450 reductase and NADH-cytochrome b reductase catalyzed the vanadate-dependent oxidation of NADPH and NADH respectively. Both of these reactions occurred aerobically and anaerobically. SOD also inhibited the V^V-dependent NAD(P)H oxidation aerobically and anaerobically. Reif et al. (1989) suggest that SOD inhibition could occur by a mechanism distinct from scavenging superoxide ion in the anaerobic experiments, such as by vanadate inhibition of SOD. Vanadate ($H_xVO_4^{n-}$) is nearly identical to phosphate ($H_xPO_4^{n-}$) structurally and phosphate is an inhibitor of SOD (Mota de Freitas et al., 1987). Clearly further investigations are required to sort out the mechanism of SOD inhibition as well as the mechanism of the vanadate-stimulated oxidation of the pyridine nucleotides.

VANADIUM(V) CATALYSIS OF HALIDE OXIDATION BY HYDROGEN PEROXIDE

Reactions of vanadium peroxide complexes with halides and their oxidized derivatives are relevant to the mechanism of the vanadium bromoperoxidase. In this section we consider the oxidation of iodide by vanadium(V)-peroxide species and the reactions of chlorine and hypochlorous acid with vanadium peroxide species.

The vanadate ion in acidic, aqueous solution promotes oxidation of iodide by hydrogen peroxide (Secco, 1980; de la Rosa and Butler, 1990). The pervanadyl ion, VO_2^+, reacts with one equivalent of hydrogen peroxide to form the monoperoxo VO_3^+ complex in acidic solution. This monoperoxo complex can react with a second equivalent of hydrogen peroxide, forming the diperoxo VO_5^- complex:

$$VO_2^+ + H_2O_2 \longrightarrow VO_3^+ + H_2O \qquad K_1 \qquad\qquad 1$$
$$VO_3^+ + H_2O_2 \longrightarrow VO_5^- + 2\,H^+ \qquad K_2 \qquad\qquad 2$$

The value of K_1 is $3.7 \times 10^4\,M^{-1}$ and independent of pH, while the value of K_2 is 0.6 M and the position of the equilibrium described by Eqn 2 is very pH dependent (Secco, 1980). Thompson's value for K_1, $3 \times 10^4\,M^{-1}$, is in agreement with the value above; however, Thompson stresses that the spectrophotometric method used to determine the formation constant is not very accurate due to the small extinction coefficients (Thompson, 1983).

Secco reports that addition of excess iodide to an equimolar solution of $NaVO_3$ and H_2O_2 in 0.05 M $HClO_4$ produces the pervanadyl ion, VO_2^+ and iodine. The rate of formation of I_2 is given by

$$d[I_2]/dt = k_a[I^-][VO_3^+] + k_b[I^-][VO_5^-] + k_c[I^-][VO_2^+] + k_d[I^-][H_2O_2]$$

The last two terms contribute negligibly to the overall rate of iodine formation because the peroxovanadium(V) species, VO_3^+, VO_5^- react much more rapidly with iodide. The mechanism of iodide oxidation is proposed in Scheme 3, reaction a-c. Reaction 3h, while not reported by Secco (1980), is required to complete the catalytic scheme. The final products of the reaction are the vanadyl ion, VO^{2+} and I_2 or I_3^-. Iodine or triiodide are formed by reactions 3f and 3g.

SCHEME 3

$$HVO_4 + I^- \longrightarrow VO_2^+ + IO^- + OH^- \qquad\qquad 3a$$
$$VO_3^+ + I^- \longrightarrow VO_2^+ + IO^- \qquad\qquad 3b$$
$$HVO_3^{2+} + I^- \longrightarrow VO_2^+ + IO^- + H^+ \qquad\qquad 3c$$

$$VO_5^- + I^- + H_2O \longrightarrow VO_3^+ + IO^- + 2\,OH^- \qquad\qquad 3d$$
$$HVO_5 + I^- \longrightarrow VO_3^+ + IO^- + OH^- \qquad\qquad 3e$$

$$IO^- + I^- + H_2O \longrightarrow I_2 + 2OH^- \qquad\qquad 3f$$
$$I_2 + I^- \longrightarrow I_3^- \qquad\qquad 3g$$

$$VO_2^+ + H_2O_2 \longrightarrow VO_3^+ + H_2O \qquad\qquad 3h$$
$$2\,VO_2^+ + 2\,I^- \longrightarrow 2\,VO^{2+} + I_2 \qquad\qquad 3i$$

Secco argues that reduction of V(V) (i.e., Rxn 3i) occurs by iodide only after all the hydrogen peroxide has been consumed. Our ^{51}V NMR data partially supports

Secco's contention since as can be seen in Figure 7, the signal area of the ^{51}V NMR resonances remains constant during the initial turnover (de la Rosa, 1990). The signal area decreases during the latter stages of the reaction probably due to reduction of VO_2^+ by iodide. Thus this system can be viewed as the *first* mimic of the non heme vanadium haloperoxidases.

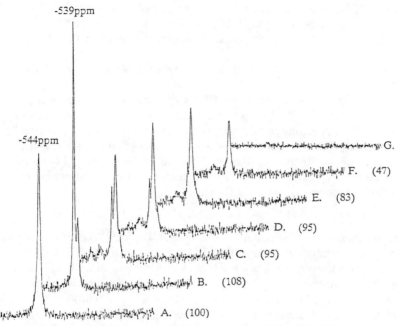

Fig. 7. ^{51}V NMR stack plot of the vanadate-catalyzed oxidation of iodide by hydrogen peroxide in 0.05 M HCLO$_4$. ^{51}V NMR spectrum of: A. 1 mM vanadate, B. 1 mM vanadate + 1 mM H$_2$O$_2$, C. 1 mM vanadate + 1 mM H$_2$O$_2$ + 10 mM NaI after 18 min, D. after 52 min, E. after 95 min., F. after 5.5 hours, G. after < 16 hours. The numbers in parentheses is the signal area in arbitrary units, relative to A. The decrease in the signal area is probably to due reduction of VO_2^+ by I$^-$. The resonance at -544 ppm arises from vanadate and the resonance at -539 ppm from oxomonoperoxovanadate.

Chlorine is an efficient oxidant of hydrogen peroxide, producing oxygen and chloride (Held *et 'al.*, 1978; Markower and Bray, 1933). Recently Thompson (1983) has investigated the oxidation of VO_3^+ by Cl$_2$:

$$VO_3^+ + Cl_2 + H_2O \longrightarrow VO_2^+ + O_2 + 2\ Cl^- + 2\ H^+$$

In the presence of excess chloride and in high acid, conditions under which the hydrolysis of Cl$_2$ is minimized, the kinetic experiments indicate that Cl$_2$ does not oxidize VO_3^+ directly. Instead the results are consistent with a mechanism involving oxidation of dissociated hydrogen peroxide, even though the ratio of uncomplexed H$_2$O$_2$ to VO_3^+ is very small (i.e., 10^{-2}–6×10^{-4}):

$$O = V^V(O_2)^+ H_2O \longrightarrow VO_2^+ + H_2O_2$$
$$H_2O_2 + Cl_2 \longrightarrow O_2 + 2\,Cl^-$$

On the other hand, preliminary results of the reaction between HOCl and VO_3^+ suggest that HOCl does oxidize VO_3^+, directly (Thompson, 1983). These results are interesting in light of vanadium(V)-bromoperoxidase which can catalyze the production of dioxygen by a bromide-assisted disproportionation of hydrogen peroxide (Everett and Butler, 1989). In fact the dioxygen that V-BrPO produces is singlet oxygen (Everett *et al.*, 1990) which may arise from reaction of H_2O_2 with HOBr, Br_2, $V_{enz}OBr$ or a brominated protein moiety (e.g., Br-amine).

VANADIUM(V) PHOTOCATALYSIS OF HYDROGEN PEROXIDE DECOMPOSITION

As previously discussed, vanadium(V) readily reacts with hydrogen peroxide to form monoperoxo and diperoxo complexes. These complexes are stable in the dark. Ultraviolet photolysis of oxoperoxovanadium(V), VO_3^+, or oxodiperoxo-vanadium(V), VO_5^-, in perchloric acid decomposes hydrogen peroxide producing dioxygen (de la Rosa and Butler, 1988; Shinohara and Nakamura, 1989). Shinohara and Nakamura (1989) have reported that the stoichiometry of the photolytic reaction in 1 M $HCLO_4$ is:

$$4\,VO_3^+ + 4\,H^+ \longrightarrow 4\,VO^{2+} + 3\,O_2 + 2\,H_2O$$

Their mechanistic investigations indicate that 1) $[VO^{2+}]$ production increases linearly with irradiation time, 2) the $[VO_3^+]$ consumed equals the $[VO^{2+}]$ produced at short irradiation times, but that at longer times, more VO_3^+ is consumed than VO^{2+} produced and 3) the quantum yield increases with the energy of irradiation: $\phi_{313nm} = 0.01$; $\phi_{366nm} = 0.0038$; $\phi_{436nm} = 0.0015$. The stoichiometry of the photochemical decomposition reaction is different in the presence of 2-propanol:

$$2\,VO_3^+ + 2\,H^+ + (CH_3)_2CHOH \longrightarrow 2\,VO^{2+} + O_2 + (CH_3)_2CO + 2\,H_2O$$

2-propanol scavenges $OH\cdot$ producing acetone. Thus, the mechanism of the photodecomposition proposed by Shinohara and Nakamura (1989) involves the formation of hydroxyl radicals in the absence of 2 propanol (Scheme 4).

SCHEME 4

$$VO(O_2)^+ \xrightarrow{\quad h\nu \quad} [O = V^{IV}-O-O\cdot{}^+]^*$$

$$[O = V^{IV}-O-O\cdot{}^+]^* + H^+ \longrightarrow VO^{2+} + HO_2 \qquad\qquad a$$

$$VO(O_2)^+ + HO_2 + H^+ \longrightarrow VO^{2+} + O_2 + H_2O_2 \qquad\qquad b$$

$$VO^{2+} + H_2O_2 \longrightarrow VO_2^+ + OH\cdot + H^+ \qquad\qquad c$$

$$VO(O_2)^+ + OH\cdot \longrightarrow VO_2^+ + HO_2 \qquad\qquad d$$

$$VO_2^+ + H_2O_2 \longrightarrow VO(O_2)^+ + H_2O \qquad\qquad e$$

This scheme (Shinohara and Nakamura, 1989) does not predict the correct stoichiometry as presented and requires additionally reaction a, reaction b twice and reaction e:

$$VO(O_2)^+ \xrightarrow{\text{hv}} [O = V^{IV}-O-O \cdot {}^+]^*$$

$$[O = V^{IV}-O-O \cdot {}^+]^* + H^+ \longrightarrow VO^{2+} + HO_2 \qquad \text{a}$$

$$VO(O_2)^+ + HO_2 + H^+ \longrightarrow VO^{2+} + O_2 + H_2O_2 \qquad \text{b}$$

$$VO(O_2)^+ + HO_2 + H^+ \longrightarrow VO^{2+} + O_2 + H_2O_2 \qquad \text{b}$$

$$VO_2^+ + H_2O_2 \longrightarrow VO(O_2)^+ + H_2O \qquad \text{e}$$

In contrast to the photochemical reactions described above which were investigated under stoichiometric vanadium/peroxide conditions, de la Rosa and Butler (1988) have found that the photochemical decomposition of hydrogen peroxide is catalyzed by a variety of vanadium(V) complexes, including $V(OQ)_2(O)(OR)$ [OQ^- = 8-hydroxyquinolinate] in alcoholic solvents (ROH) such as methanol and ethanol, and V(V)-desferrioxamine or $H_2VO_4^-$ in aqueous solution at neutral pH. The reactions, which have been followed spectrophotometrically and by [51]V NMR, demonstrate that the vanadium(V) complex is regenerated after consumption of all the hydrogen peroxide. Figure 8 shows the [51]V NMR spectra of the photolysis of vanadium(V) quinolinato-peroxy complex in methanol.

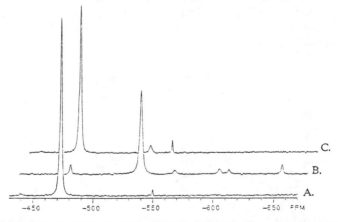

Fig. 8. [51]V NMR stack plot of the photolysis of $V(OQ)_2(O)(O_2)$ [OQ^- = 8-hydroxyquinolinate]. A. $V(OQ)_2(O)(MeO)$ in methanol, B. $V(OQ)_2(O)(O_2)$, C. $V(OQ)_2(O)(O_2)$ + photolysis.

NUCLEASE ACTIVITY OF VANADYL BLEOMYCIN

The bleomycins are a group of glycopeptides that bind to and cleave DNA in the presence of dioxygen (or a reduced species, e.g., hydrogen peroxide) and a metal ion (e.g., iron) (for a review see Stubbe and Kozarich, 1987). Vanadyl ion also forms a 1:1, orange complex with bleomycin between pH 6 and 10 (Kuwahara *et al.*, 1985). The uv-vis spectrum has absorption maxima at 467 nm and 758 nm which have been assigned to dxy-dx^2-y^2 and dxy-dxz,yz ligand field transitions. The ESR spectrum is consistent with square pyramidal coordination of VO^{2+} by five nitrogen ligands, producing a six coordinate complex. The proposed ligands are a secondary amine nitrogen, the pyrimidine nitrogen, a deprotonated amide nitrogen, histidyl imidazole nitrogen and an α-amine nitrogen as an axial ligand, consistent with the proposed ligands for other metal bleomycin complexes. The ESR signal disappears upon addition of hydrogen peroxide, which presumably oxidizes V(IV) to V(V); addition of dithionite restores the original ESR signal.

VO^{2+}-bleomycin catalyzes double stranded DNA (i.e., supercoiled pBR 322) scission in the presence of excess hydrogen peroxide, forming nicked and linear DNA. Vanadyl-bleomycin is about 50 times less effective than iron-bleomycin, under the conditions employed. Nucleotides adjacent to guanine bases such as G-C and G-A (5' → 3') were preferentially cleaved by the vanadium bleomycin derivative. In addition, vanadium-bleomycin attacked G-A (5' → 3') sequences more frequently than the iron-bleomycin complex. The authors suggest that the different reactivities of the vanadyl and ferric bleomycins might be due to different configurations.

OXYGENATION REACTIVITY OF VANADIUM(V) PEROXO COMPLEXES

The industrial process for the synthesis of propylene oxide utilizes d° metal complexes of V(V), Mo(VI) or Ti(IV) to catalyze the oxygen-atom transfer from alkyl hydroperoxide to propylene. Other examples of oxidation reactions effected by V(V)-peroxide and V(V)-alkyl peroxide complexes are also well known (for a

V(V)(R'-OPhsal-R″)

review see Sharpless and Verhoeven, 1979). Two examples are highlighted herein. The vanadium(V) alkylperoxy complexes of N-(2-oxidophenyl)-salicylidenamine (R'-OPhsal-R") (R', R"H = CH$_3$, Cl, NO$_2$; see structure below) epoxidize olefins stereospecifically and in high yield (Mimoun et al., 1986).

The structure of VO(OOR)(R'-OPhsal-R") is probably pentagonal bipyramidal, consistent with almost all other peroxoV(V) complexes characterized x-ray crystallographically. Epoxidation reactions catalyzed by VO(OOtBu)(R'-OPhsal-R") are completely stereoselective, giving consistently high yields (48-98%) in anhydrous solvents at room temperature. Cis and trans 2,3-epoxybutane were produced from cis 2-butene and trans 2-butene, respectively in 65-70% yield and detection of other products was not reported (Mimoun et al., 1986). Exoepoxynorbornane was the sole product produced from norbornene. Styrene is epoxidized, exclusively, however, styrene epoxide is oxidized by another equivalent of VO(OOtBu)(R'-OPhsal-R") producing benzaldehyde. Allylic alcohol is not reactive. These reactions are inhibited by water, alcohols, basic ligand and solvents and accelerated in polar, non donar solvents.

The epoxidation results above were obtained under stoichiometric V^V-peroxide conditions. A catalytic mechanism of epoxidation catalyzed by VO(OOtBu)-(R'-OPhsal-R"), as proposed by Mimoun et al. (1986) is presented in Scheme 5.

SCHEME 5

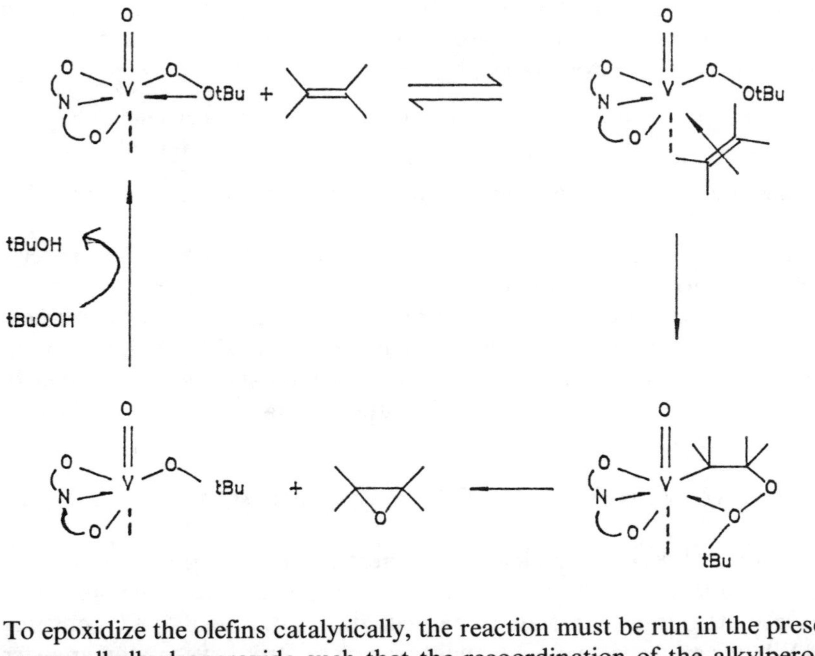

To epoxidize the olefins catalytically, the reaction must be run in the presence of excess alkylhydroperoxide such that the recoordination of the alkylperoxide to VO(OtBu)(R'-OPhsal-R") can occur. The kinetics of epoxidation exhibit saturation behavior upon increasing the concentration of the olefin, suggesting that the olefin reversibly forms a complex with the V(V)-peroxo complex before oxygenation occurs. Insertion of the olefin into the V-O bond of the V-OOtBu, the hypothesized dioxametallocycle is the rate limiting step under the conditions employed. Oxygen-atom transfer occurs rapidly by a heterolytic cleavage reaction, producing the epoxide. Vanadium(V) functions as a Lewis acid.

The *tert*-butyl peroxide VO^{3+} complex of dipicolinate, on the other hand, does not epoxidize olefins. The reactivity difference between the dipicolinate and R'-OPhSal-R″complexes has been ascribed to the lack of a coordination site coplanar and adjacent to the alkyl peroxide for the olefin in the dipicolinate complex. This may arise from strong coordination of the *tert*-butoxy oxygen atom in the dipicolinate complex but weak coordination of the *tert*-butoxy oxygen atom in the R'-OPhSal-R″ complex.

In contrast to the stereoselective reactivity of VO(OOtBu)(R'-OPhsal-R″), Mimoun *et al.* (1983) have reported that V(V) oxoperoxo complexes of the general formula $(VO(O_2)(pic)LL'$ (pic = pyridine-2-carboxylate; L,L' = H_2O, MeOH, monodendate and bidentate ligands) epoxidize and oxidize olefins nonstereo-specifically (i.e., *cis*-2-butene produces a mixture of cis and trans 2-epoxybutane). More interestingly, however, the $VO(O_2)(pic)LL'$ complexes hydroxylate aromatic hydrocarbons, forming phenols and hydroxylate alkanes, forming alcohols and ketones (Mimoun *et al.*, 1983).

The $VO(O_2)(pic)LL'$ complexes are reactive only in aprotic solvents. These solvents also promote the decomposition of the peroxo complex, forming a dimeric *bis* μ-oxo species. This product is also formed during the oxidation of olefins, alkanes and aromatic hydrocarbons:

$$(pic)V^V(O)(O_2)LL' + \text{olefin, alkane, aromatic hydrocarbon}$$
$$\longrightarrow (pic)V(=O)(\mu-O)_2V(=O)(pic) + \text{O-alkane}$$

In all of these reactions, the transferred oxygen atom originates from the peroxide (Mimoun, 1983; 1986).

Mimoun has ascribed the reactive oxidant to the diradical V^{IV}-OO· species based on a comparison of reactivity of peroxidic reagents like the R_2C·-OO·: 1) the lack of stereospecificity in the epoxidation is reminiscent of alkylperoxy radical reactions, 2) the hydroxylation of alkanes can be explained by hydrogen-atom abstraction by V^{IV}-OO·, and 3) hydroxylation of aromatics is consistent with addition of V^{IV}-OO· to the ring and not H-atom abstraction due to the lack of a primary deuterium isotope effect (Mimoun, 1983).

VANADIUM-OXYGEN ATOM TRANSFER AND OXIDE DISPLACEMENT REACTIONS

Recently Holm (1987) has reviewed the literature on oxygen-atom transfer reactions, including vanadium-oxo moieties. While vanadium has the potential to effect oxygen-atom transfer reactions, particularly V^V=O and V^{IV}=O producing V^{III} and V^{II} species, respectively, not many examples of true oxygen-atom transfer reactivity are known for vanadium. For further discussion of oxygen-atom transfer reactions of vanadium, the reader is referred to Holm's review (Holm, 1987).

The SALEN (N,N'-ethylenebis-{salicylideneamine}) and SALDPT (N,N'-(3,3'dipropylamine)-bis(salicylideneamine)) complexes of vanadium (III, IV) have been found to be effective oxygenation catalysts of 3,5 di-t-Bu-pyrocatechol using dioxygen (Tatsuno *et al.*, 1982). The products are 3,5-di-t-butylmuconic acid anhydride, principally, and also 4,6-di-t-butyl-2-pyrone and 3,5-di-t-butyl-o-benzoquinone. Compared to other catalytic systems that carry out this reaction, including pyrocatechase, these vanadium complexes appear to be intra-diol cleavage catalysts, primarily.

VANADIUM(IV)-OXIDE DISPLACEMENT AND DISPROPORTIONATION

The oxide of vanadyl-SALEN can be displaced by reaction with oxophilic reagents such as thionyl chloride ($SOCl_2$) (Pasquali *et al.*, 1979), Ph_3PBr_2 (Callahan and Durand, 1980) as well as anhydrous HCl in acetonitrile (Bonadies *et al.*, 1987), forming the bare V^{4+} ion:

$$O{=}V^{IV}SALEN + SOCl_2 \longrightarrow V^{IV}Cl_2SALEN + SO_2 + H_2O$$

The oxide can also be displaced by sulfides (e.g., B_2S_3, Callahan and Durand, 1980; $(Me_3Si)_2S$, Do *et al.*, 1983) forming thiovanadyl species:

$$O{=}V^{IV}SALEN + (Me_3Si)_2S \longrightarrow S{=}V^{IV}SALEN + (Me_3Si)_2O$$

VO^{2+} SALEN undergoes a novel disproportionation reaction upon addition of certain acids in acetonitrile (Bonadies *et al.*, 1987). Addition of "non coordinating" anions such as perchloric acid, hydrofluoric acid (or sulfuric acid, Pecoraro, personal communication) to an anaerobic solution of VO^{2+} SALEN in acetonitrile produces a green-brown solution containing equimolar mixture of $V^VO(SALEN)ClO_4$ and a V(III) species. This reaction is accompanied by the loss of ESR signal of $VO(SALEN)^{2+}$. If half the vanadium concentration is assumed to be in the diamagnetic V(V) oxidation state, the resultant solution susceptibility is consistent with V(III), i.e., $\mu = 3.1$. The UV-Vis spectrum of the solution is also consistent with V(V)OSALEN and V(III)SALEN species. Addition of chloride induces the conproportionation of this V(V)OSALEN/V(III)SALEN mixture, forming a mixture of $V^{IV}O(SALEN)$ and $V^{IV}Cl_2(SALEN)$. Chloride apparently stabilizes the V(IV) oxidation state. Pecoraro and Carrano have suggested that this novel disproportionation reaction may be a route to the formation and accumulation of V(III) in the tunicates (Bonadies *et al.*, 1986; 1987).

VANADIUM-ASSISTED OXIDATIVE DECARBOXYLATION OF ETHYLENEBIS((O-HYDROXYPHENYL)GLYCINE)

The V(V)complex of ethylenebis((o-hydroxyphenyl)glycine) (EHPG) forms a dark blue monooxoV(V) complex with vanadium(V), $V^VO(EHPG)$ (Pecoraro *et al.*, 1984). In this complex VO^{3+} is coordinated to EHPG via two phenolates, two ethylenediamine nitrogens and a carboxylate ligand. The other carboxylate is not coordinated and exists as a pendant protonated moiety.

V^{V}O-EHPG

In warm acetone or DMF, V^{V}O(EHPG) is converted to VO^{2+}(SALEN). The complete conversion of V^{V}O(EHPG) to VO^{2+}(SALEN) requires the presence of dioxygen and effects the loss of two equivalents of carbon dioxide, three protons and three electrons per V^{V}O(EHPG) complex and the oxidation of the two ethylenediamine amine nitrogens to imino nitrogens.

In alcoholic solvents, in which the oxidative decarboxylation occurs slowly, a monodecarboxylated, monoimine, monooxo vanadium(V) compound of N,N'-ethylene((o-hydroxyphenyl)-glycine)salicylidenimine (EHGS) was isolated (Pecoraro *et al.*, 1985). VO(EHGS) is also a dark blue compound which can be reversibly reduced forming the orange V^{IV}O(EHGS) species. In DMF V^{V}O(EHGS) is converted to V^{IV}O(SALEN) very rapidly. For this reason, the V^{IV,V}O(EHGS) species are not observed as intermediates in the successive oxidative decarboxylation of V^{V}O(EHPG) to VO^{2+}(SALEN) in DMF. The mechanism of this novel vanadium-catalyzed decarboxylation of EHPG producing EHGS is proposed in Scheme 6.

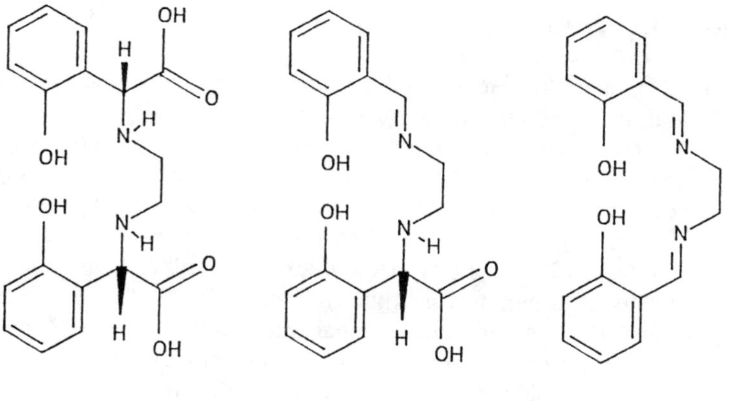

EHPG   EHGS   SALEN

SCHEME 6

Proposed mechanism for the first decarboxylation of EHPG

$$V^VO(EHPG) \longrightarrow V^{III}(EHGS) + CO_2$$
$$V^VO(EHPG) + V^{III}(EHGS) \longrightarrow V^{IV}O(EHGS) + V^{IV}O(EHPG)$$
$$V^{IV}O(EHGS) \xrightarrow{\;O_2\;} V^VO(EHGS)$$
$$V^{IV}O(EHPG) \xrightarrow{\;O_2\;} V^VO(EHPG)$$

A key feature of the mechanism is the two electron reduction of V(V) which occurs in the decarboxylation step. This feature is supported by the result that heating $V^VO(EHPG)$ in the absence of dioxygen produces $V^{III}(EHGS)$ and CO_2 (Bonadies and Carrano, 1985). The conversion of EHGS to SALEN, i.e., the second decarboxylation, probably occurs in a manner analogous to that presented in Scheme 6. The mechanism of vanadium catalyzed decarboxylation of EHPG is further supported by the isolation and X-ray crystallographic characterization of many of the important V(IV) and V(V) derivatives of EHPG (Riley *et al.*, 1986), shown in Scheme 6.

Recently vanadium(V) has been used to carry out the oxidative decarboxylation of 3-hydroxycarboxylic acids which occurs upon stoichiometric addition of $VOCl_3$ (Meier and Schwartz, 1989). These authors also report the formation of olefins, which until now has been unprecedented in the presence of an oxophilic metal species.

V(II) PHOTOCHEMISTRY

Vanadium(II) phenanthroline and polypyridine complexes are being investigated as photocatalysts of multielectron oxidations. Maverick has shown that the electronic excited state of *tris*-phenanthroline vanadium(II), $V(phen)_3^{2+}$, and *tris*-bipyridine vanadium(II), $V(bpy)_3^{2+}$, are oxidatively quenched by appropriate electron acceptors. The lowest energy excitation is a metal to ligand charge transfer state (i.e, $^4A_2 \longleftarrow {}^4MLCT$) that lies slightly lower in energy than the ligand field states, $^4T2$, ^2E and $^2T_1$ (Maverick *et al.*, 1987; Konig and Herzog, 1970). The excited state lifetimes of $V(phen)_3^{2+}$ and $V(bpy)_3^{2+}$ in EtOH are short, i.e., 1.8 nsec and 0.5 nsec, respectively (Shah and Maverick, 1986). Nevertheless, photolysis of $V(phen)_3^{2+}$ at 640 nm in the presence of an electron acceptor, e.g., methyl viologen, europium(II) or $VO(phen)_2^{2+}$, produces a binuclear V(III) dimer, $[V_2(phen)_4(\mu\text{-}OH)_4]^{4+}$ (Scheme 7).

SCHEME 7

$V(phen)_3^{2+}$ $[^4A_2] \longrightarrow V(phen)_3^{2+*}$ $[^4MLCT]$

$V(phen)_3^{2+*}$ $[^4MLCT]$ + Acceptor $\longrightarrow V(phen)_3^{3+}$ + Acceptor^{red-}

$2V(phen)_3^{3+}$ + $2H_2O \longrightarrow [(phen)_2V(\mu\text{-}OH)_2V(phen)_2]^{4+}$ + 2phen + $2H^+$

The back electron transfer reaction is inhibited by the hydrolysis and subsequent dimerization of $V(phen)_3^+$. The dimer can be further oxidized in basic solution (i.e., pH <8) by methyl viologen, 1,1'-dimethyl-4,4'-bipyridinium and $Fe(CN)_6^{3-}$, producing VO^{2+}, resulting in a net two electron photooxidation of $V(phen)_3^{2+}$ (Shah and Maverick, 1987). $V(bpy)_3^{2+}$ reacts by an analogous reaction scheme (Shah and Maverick, 1986).

Another route to the formation of the VO^{2+} state is by oxidation of the vanadium(II) pyridyl complexes by dioxygen (Bennett and Taube, 1968; Dobson and Taube, 1989). Oxidation of $V(trpy)(bpy)(H_2O)^{2+}$ (trpy = 2,2':6',2"-terpyridine) by dioxygen produces $V^{IV}(trpy)(bpy)O^{2+}$, provided the concentration of the vanadium is low, i.e., $<10^{-4}$ M (Dobson and Taube, 1989). At higher concentrations, formation of the V(III)-dimer, (trpy)(bpy)V^{III}-O-V^{III}(bpy)trpy)$^{4+}$, inhibits formation of the vanadyl species, presumably do to reaction of $V^{IV}(trpy)(bpy)O^{2+}$ with unoxidized $V(trpy)(bpy)(H_2O)^{2+}$. $V^{IV}(trpy)(bpy)O^{2+}$ does not oxidize the organic compounds, 2-propanol, styrene nor triphenylphosphene even at elevated temperatures for long periods of time (Dobson and Taube, 1989). On the other hand, $V^{IV}(trpy)(bpy)O^{2+}$ is further oxidized to $V^V(terpy)(byp)$ complexes, which have not been well characterized. Such V(V) complexes are interesting in light of the coordination of strongly π-accepting pyridyl ligands to a d^0 metal center (Dobson and Taube, 1989). Their characterization and reactivity should be quite interesting.

VANADIUM(II) REDUCTION OF DINITROGEN

In 1970, Shilov *et al.* (1970) reported that basic solutions of vanadium(II) and magnesium(II) were capable of reducing dinitrogen to hydrazine (N_2H_4) as well as acetylene to ethylene and ethylene to ethane (for a review see Shilov, 1980). A tremendous debate over the mechanism of the reduction has ensued. Schrauzer *et al.* have proposed that the principal step is the formation of diimide (N_2H_2) in a two electron reduction of dinitrogen by V(II). Shilov, on the other hand, have proposed that V(II) acts as a one-electron reductant and has invoked a multinuclear V(II)/Mg(II) complex. For further discussion, the reader is referred to Boas and Pessoa, 1987 and to reviews by Schrauzer (1980) and Shilov (1980).

Acknowledgements

AB gratefully acknowledges grants from the National Institutes of Health (GM-38130), the National Science Foundation (DMB87-16229), the donors of the Petroleum Research Fund, administered by the American Chemical Society, (PRF # 17895) and the American Cancer Society, Junior Faculty Research Award (JFRA-216). The unpublished work on the V-BrPO mimics and V-peroxide photoreactions followed by ^{51}V NMR was supported by NIH.

References

Beas, C.F., Mesmer, R.E.: 1976, The Hydrolysis of Cations, Wiley-Interscience, NY, 197–210.

Bennett, L.E., Taube, H.: 1968, *Inorg. Chem.* **7**, 254.

Boas, L.V., Pessoa, J.C.: 1987, in Comprehensive Coordination Chemistry, The Synthesis, Reactions, Properties and Applications of Coordination Compounds, Ed. G. Wilkinson, Pergamon Press, Volume 3, 453–583.

Bonadies, J.A., Carrano, C.J.: 1986, *J. Am. Chem. Soc.* **108**, 4088–4095.

Bonadies, J.A., Pecoraro, V.L., Carrano, C.J.: 1986, *J. Chem. Soc., Chem. Commun.*, 1218.

Bonadies, J.A., Butler, W.M., Pecoraro, V.L., Carrano, C.J.: 1987, *Inorg. Chem.* **26**, 1218–1222.

Bonadies, J.A., Carrano, C.J.: 1986, *Inorg. Chem.* **25**, 4358–4361.

Boyd, D.W., Kustin, K.: 1984, *Advances in Inorganic Biochemistry* **6**, 311–365.

Butler, A., Eckert, H.: 1987, *J. Amer. Chem. Soc.* **109**, 1864–1865.

Butler, A., Eckert, H.: 1989, *J. Amer. Chem. Soc.* **111**, 2802–2809.

Butler, A., Parsons, S.M., Yamagata, S.K., de la Rosa, R.I.: 1989, *Inorganica Chimica Acta* **163**, 1–3.

Callahan, K.P., Durand, P,.J.: 1980, *Inorg. Chem.* **19**, 3211.

Chasteen, N.D.: 1981, *Biol. Mag. Res.* **3**, 53–119.

Chasteen, N.D.: 1983, *Structure and Bonding* **53**, 105–138.

Ciurli, A., Holm, R.H.: 1989, *Inorg. Chem.* **28**, 1685–1690.

Comba, P., Engelhardt, L.M., Harrowfield, J.MacB., Lawrance, G.A., Martin, L.L., Sargeson, A.M., White, A.H.: 1985, *J. Chem. Soc. Chem Commun.*, 174–176.

Comba, P., Sargeson, A.M.: 1986, *Aust. J. Chem.* **39**, 1029–1033.

Coulombe, R.A., Briskin, D.P., Keller, R.J., Thornley, W.R., Sharma, R.P.: 1987, *Arch. Biochem. Biophys.* **255**, 267–273.

Crans, D.C., Shin, P.K.: 1988, *Inorg. Chem.* **27**, 1797–1806.

Darr, D., Fridovich, I.: 1984, *Arch. Biochem. Biophys.* **232**, 562–565.

Darr, D., Fridovich, I.: 1985, *Arch. Biochem. Biophys.* **243**, 220–227.

de la Rosa, R.I., Butler, A.: Abstracts of the LA ACS meeting, 1988, Los Angeles, CA.

de la Rosa, R.I., Butler, A.: Manuscript in Preparation.

Desideri, A., Raynor, J.B., Diamantis, A.A.: 1978, *J. Chem. Soc. Dalton*, 423–426

Diamantis, A.A., Raynor, J.B., Rieger, P.H.: 1980, *J. Chem. Soc. Dalton*, 1731–1733.

Djordjevic, C., Craig, S.A., Sinn, E.: 1985, *Inorg. Chem.* **24**, 1281–1283.

Djordjevic, C., Puryear, B.C., Vuletic, N., Abelt, C.J., Sheffield, S.J.: 1988, *Inorg. Chem.*, 2926–2932.

Djordjevic, C., Lee, M., Sinn, E.: 1989, *Inorg. Chem.* **28**, 719–723.

Do, Y., Simhon, E.D., Holm, R.H.: 1983, *Inorg. Chem.* **22**, 3809.

Dobson, J.C., Taube, H.: 1989, *Inorg. Chem.* **28**, 1310–1315.

Drew, R.E., Einstein, F.W.B.: 1972, *Inorg. Chem.* **11**, 1079–1083.

Everett, R.R., Butler, A.: 1989, *Inorganic Chemistry*, 28, 393–395.

Everett, R.R., Kanofsky, J.R., Alison Butler, A.: 1990, *J. Biol. Chem.* **265**, in press.

Funahashi, S., Midorikawa, T., Tanaka, M.: 1980, *Inorg. Chem.* **19**, 91–97.

Gambarotta, S., Mazzanti, M., Floriani, C., Chiesi-Villa, A., Guastini, C.: 1985, *J. Chem. Soc. Chem. Commun.*, 829–830.

Giacomelli, A., Floriani, C., Duarte, A,O dS., Chiesi-Villa, A., Guastini, C.: 1982, *Inorg. Chem.* **21**, 3310–3316.

Halbert, T.R., Hutchings, L.L., Rhodes, R., Stiefel, E.I.: 1986, *J. Am. Chem. Soc.* **108**, 6437–6438.

Hambley, T.W., Hawkins, C.J., Kabanos, T.A.: 1987, *Inorg. Chem.* **26**, 3740.

Hawkins, C.J., Kabanos, T.A.: 1989, *Inorg. Chem.* **28**, 1084–1087.

Hayasi, Y., Schwartz, J.: 1981, *Inorg. Chem.* **20**, 3473–3476.

Held, A.M., Halko, D.J., Hurst, J.K.; 1978, *J. Am. Chem. Soc.* **101**, 5732.

Holm, R.H.: 1987, *Chem. Rev.* **87**, 1401–1449.

Isreal, Y., Meites, L.: 1985 in Standard Potentials in Aqueous Solution, eds A.J. Bard, R. Parsons and J.Jordan, Dekker, New York, 1985.

Khandke, L., Gullapalli, S., Patole, M.S., Ramasarma, T.: 1986, *Arch. Biochem. BIophys.* **244**, 742–749.

Konig, E., Herzog, S.; 1970, *J. Inorg. Nucl. Chem.* **32**, 601.

Kovacs, J.A., Holm, R.H.: 1987a, *Inorg. Chem.* **26** 702–711.

Kovacs, J.A., Holm, R.H.: 1986, *J. Am. Chem. Soc.* **108**, 340–341.

Kovacs, J.A., Holm, R.H.: 1987b, *Inorg. Chem.* **26** 711–717.

Kustin, K., Macara, I.G.: 1982, *Comm. Inorg. Chem.* **2**, 1–22.

Kustin, K., McLeod, G.C., Gilbert, T.R., Briggs, L B.R., 4th.: 1983, *Structure and Bonding* **53**, 139–160.

Kuwahara, J., Suzuki, T., Sugiura, Y., 1985, *Biochem. Biophys. Res. Commun.* **129**, 368–374.

Lah, M.S., Pecoraro, V.L., Kirk, M.L., Hatfield, W.: 1990, in Biosensors, R. Buck, E. Bowden, M. Umana Eds., Marcel Dekker, NY in press.

Latimer, W.M.: 1952, The Oxidation States of the Elements and Their Potentials in Aqueous Solutions, 2nd Ed., Prentice-Hall, Inc., NY.

Liochev, S., Ivancheva, E., Fridovich, I.: 1989, *Arch. Biochem. Biophys.* **269**, 188–193.

Liochev, S., Fridovich, I.: 1988, *Arch. Biochem. Biophys.* **263**, 299–304.

Liochev, S., Fridovich, I.: 1987, *Arch. Biochem. Biophys.* **255**, 274–278.

Markower, B., Bray, W.C.: 1933, *J. Am. Chem. Soc.* **55**, 4765–0000.

Matsubayashi, D-e., Akiba, K., Tanaka, T.: 1988, *Inorg. Chem.* 4744–4749.

Maverick, A.W., Shah, S.S., Kirmaier, C., Holton, D.: 1987, *Inorg. Chem.* **26**, 774–776.

Mazzanti, M., Gambarotta, S., Floriani, C., Chiesi-Villa, A., Guastini, C.: 1986, *Inorg. Chem.* **25**, 2308–2314.

Meier, I.K., Schwartz, J.: 1989, *J. Am. Chem. Soc.* **111**, 3069–3070.

Mimoun, H., Saussine, L., Daire, E., Postel, M., Fischer, J., Weiss, R.: 1983, *J. Am. Chem. Soc.* **105**, 3101–3110.

Mimoun, H., Mignard, M., Brechot, P., Saussine, L.: 1986, *J. Am. Chem. Soc.* **108**, 3711–3718.

Mota de Freitas, D., Luchinat, C., Banci, L., Bertini, I., Valentine, J.S.: 1987, *Inorg. Chem.* **26**, 2788–2791.

Nechay, B.R.: 1984, *Ann. Rev. Pharmacol. Toxicol.* **24**, 501–524.

Orhanovic, M., Wilkins, R.G.: 1967, *J. Am. Chem. Soc.* **89**, 278–282.

Pasquali, M., Marchetti, F., Floriani, C.: 1979, *Inorg. Chem.* **18**, 2401–2404.

Patole, M.S., Kurup, C.K.R., Ramasarma, T.: 1986, *Biochem Biophys. Res. Commun.* **141**, 171–175.

Pecoraro, V., Bonadies, J.A., Marrese, C.A., Carrano, C.J.: 1984, *J. Am. Chem. Soc.* **106**, 3360–3362.

Pecoraro, V.L.: 1989, *Inorg. Chim. Acta* **155**, 171–173.

Ramasara, T., Crane, F.L.: 1981, *Curr. Topics Cell Reg.* **20**, 247–301.

Ramasarma, T., MacKellar, W.C., Crane, F.L.: 1981, *Biochem. Biophys. Acta* **646**, 88–98.

Reif, D.W., Coulombe, Jr., R.A., Aust, S.: 1989, *Arch. Biochem. Biophys.* **270**, 137–143.

Riley, P.E., Pecoraro, V.L., Carrano, C.J., Bonadies, J.A., Raymond, K.N.: 1986, *Inorg. Chem.* **25**, 154–160.

Rush, J.D., Bielski, B.H.J.: 1985, *J. Phys. Chem.* **89**, 1524–1528.

Saito, K., Sasaki, Y.: 1988, *Pure & Appl. Chem.* **60**, 1123–1132.

Schrauzer, G.N.: 1980, in New Trends in the Chemistry of Nitrogen Fixation, ed. J. Chatt, L.M.C. Pina, R.L. Richards, Academic Press, London, 103–121.

Secco, F.: 1980, *Inorg. Chem.* **19**, 2722–2725.

Shah, S.S., Maverick, A.W., 1987, *Inorg. Chem.* **26**, 1559–1562.

Shah, S.S., Maverick, A.W.: 1986, *Inorg. Chem.* **25**, 1867–1871.

Sharpless, K.B., Verhoeven, T.R.: 1979, Aldrichimica Acta **12**, 63–74.

Shilov, A.E.: 1980 in New Trends in the Chemistry of Nitrogen Fixation, ed. J. Chatt, L.M.C. Pina, R.L. Richards, Academic Press, London, 121.

Shinohara, N., Nakamura, Y.: 1989, *Bull. Chem. Soc. Jpn.* **62**, 734–737.

Stiefel, E.I., Dori, A., Gray, H.B.: 1967, *J. Am. Chem. Soc.* **89**, 3353–3354.

Stomberg, R.: 1986, *Acta Chem. Scand.* A**40**, 168–176.

Stubbe, J., Kozarich, J.W.: 1987, *Chemical Reviews* **87**, 1107–1136.

Szentivanyi, H., Stomberg, R.: 1983, *Acta Chem. Scand.* A**37**, 709–714.

Tatsuno, Y., Tatsuda, M., Otsuka, S.: 1982, *J. Chem. Soc., Chem Commun.*, 1100–1101.

Thompson, R.C.: 1983, *Inorg. Chem.* **22**, 584–588.

Toma, H.E., Santos, P.S., Lellis, F.T.P.: 1988, *J. Coord. Chem.* **18**, 307–316.

Tracey, A.S., Gresser, M.J., Galeffi, B.: 1988a, *Inorg. Chem.* **27**, 157–161.

Tracey, A.S., Gresser, M.J.: 1988b, *Inorg. Chem.* **27**, 1269–1275.

Welch, J.H., Bereman, R.D., Singh, P.: 1988, *Inorg. Chem.* **27**, 2862–2863.

Wieghardt, K.: 1978, *Inorg. Chem.* **17**, 57–64.

III. The Essentiality and Metabolism of Vanadium

FORREST H. NIELSEN and ERIC O. UTHUS
United States Department of Agriculture, Grand Forks Human Nutrition Research Center,
P.O. Box 7166, University Station, Grand Forks, ND 58202, U.S.A.

The hypothesis that vanadium is nutritionally essential for higher animals, including humans, has had a long and checkered history. In 1949, Rygh (1949) reported that vanadium might be needed by animals because vanadium stimulated the mineralization of bones and teeth, and prevented caries formation in rats and guinea pigs. However, Bertrand (1950) stated in 1950 that "we are completely ignorant of the physiological role of vanadium in animals, where its presence is constant." Schroeder *et al.* (1963) reviewed the literature up to 1963 and concluded that, although vanadium behaves like an essential trace element, final proof of essentiality for mammals was still lacking. Between 1971 and 1974, a number of findings reported by several different research groups (Schwarz and Milne, 1971; Strasia, 1971; Hopkins and Mohr, 1971, 1974; Nielsen and Ollerich, 1973; Williams, 1973; Nielsen and Sandstead, 1974) led many to conclude that vanadium is an essential nutrient. However, subsequent reports (Nielsen, 1985a; Nechay *et al.*, 1986) presented a convincing argument that most of the evidence presented as proof for the essentiality of vanadium was nothing more than evidence that vanadium has very active in vitro and pharmacologic actions. Only recently the conclusion of Schroeder *et al.* (1963) that "no other trace metal has so long had so many supposed biological activities without having been proved to be essential" has begun to crumble. Recent indication of apparent vanadium deficiency signs in goats (Anke *et al.*, 1986) and rats (Uthus and Nielsen, 1988, 1989) and the discovery of vanadium-containing enzymes (Vilter, 1984; De Boer *et al.*, 1986a, 1986b; Krenn *et al.*, 1987; Plat *et al.*, 1987; Vilter and Rehder, 1987) in plants and microorganisms support the concept that vanadium is an essential nutrient for higher animals. Thus, it seems appropriate to separate some of the reports on the nutritional and metabolic aspects of vanadium from the voluminous literature on the element and to review them here.

1. Essentiality

Before discussing the essentiality of vanadium, it seems appropriate to give a definition of the term. A substance is usually considered nutritionally essential if a dietary deficiency consistently results in a suboptimal biological function that is preventable or reversible by physiological amounts of the substance. In this definition, physiological is construed as those quantities usually found in biological material. For vanadium, the usual amounts are measured in nanograms per gram

N. Dennis Chasteen (ed.), Vanadium in Biological Systems, 51–62.
© 1990 *Kluwer Academic Publishers. Printed in the Netherlands.*

of tissue. Another term that should be defined here is "pharmacological action." In the following discussion it means the effect of a relatively high dietary intake of a substance that either alleviates an abnormality caused by something other than a nutritional deficiency of that substance or alters some biochemical function or biological structure in a manner that may be construed as beneficial. For vanadium, pharmacological amounts probably can be as low as a microgram per gram of diet.

DEFICIENCY SIGNS

Between 1971 and 1974 four research groups described possible signs of vanadium deficiency. However, close examination of their findings revealed confusion and inconsistency. For example, in studies of rats, findings were as follows. In 1971, Strasia (1971) reported that rats fed less than 100 ng vanadium/g diet exhibited slower growth, higher plasma and bone iron, and higher hematocrits than controls fed 500 ng vanadium/g diet. Williams (1973), however, could not duplicate those findings in the same laboratory. Schwarz and Milne (1971) reported that a vanadium supplement of 0.25 to 0.50 μg/g diet gave a positive growth response in suboptimally growing rats fed a purified diet with an unknown vanadium content. On the other hand, Hopkins and Mohr (1974) reported that the only effect on rats of vanadium deprivation was impaired reproductive performance (decreased fertility and increased perinatal mortality) that became apparent only in the fourth generation.

Deprivation studies with chicks also gave inconsistent findings. Hopkins and Mohr (1971, 1974) found that vanadium-deprived chicks exhibited significantly depressed wing-and tail-feather development, depressed plasma cholesterol and triglycerides at age 28 days, and elevated plasma cholesterol at age 49 days. Nielsen and Ollerich (1973) reported that vanadium deprivation depressed growth, elevated hematocrits and plasma cholesterol, and adversely affected bone development in chicks.

Subsequent to these studies, Nielsen and co-workers (Myron et al., 1975; Nielsen and Myron, 1976; Nielsen and Uthus, 1977; Nielsen et al., 1978; Uthus et al., 1978; Nielsen, 1980) tried to establish a reproducible set of signs of vanadium deprivation for both chickens and rats. In several experiments with rats fed diets of different composition, vanadium deprivation adversely affected perinatal survival, growth, physical appearance, hematocrit, plasma cholesterol, and lipids and phospholipids in the liver. Unfortunately, no sign of deprivation was found consistently throughout all experiments. Findings from studies of chicks were similarly inconsistent.

In the preceding vanadium deprivation studies, "controls" or supplemented animals were fed 0.5 to 3.0 μg vanadium/g diet. These doses of available vanadium apparently are 10 to 100 times those normally found in a diet under natural conditions. Vanadium is a relatively toxic and pharmacologically active element. For example, high vanadium can control the high blood glucose and prevent the decline in cardiac performance in rats made diabetic with streptozotocin (Heyliger et al., 1985). Thus, the amounts of vanadium used in the supplemental diets most likely could exert a pharmacologic effect, especially if the experimental animal was suboptimal in a nutrient affected by vanadium. High dietary vanadium can influence the metabolism of a number of nutrients, and vice versa. Among the nutrients which interact with vanadium are chloride (Hill, 1985), iodide (Uthus and Nielsen, 1988, 1989), chromium (Hill, 1979a), iron (Nielsen et al., 1983; Nielsen, 1985b; Blalock and Hill, 1987), copper (Nielsen, 1984a; Shuler and Nielsen, 1985), ascorbic acid (Hill, 1979b), cystine (Shuler and Nielsen, 1983, 1985; Nielsen et al., 1984), methionine (Nielsen, 1985b; Shuler and Nielsen, 1985), riboflavin (Hill, 1988), and protein (level and source) (Hill, 1979c).

Most of the studies performed from 1971 to 1974 were with animals fed unbalanced diets, thus resulting in suboptimal performance such as depressed growth. The diets used often had widely varying contents of protein, sulfur-containing amino acids, ascorbic acid, iron, copper, and perhaps other nutrients that affect, or are affected by, vanadium metabolism. The following will show some of the variation in treatments in the early studies. Strasia (1971), who found that vanadium affected iron metabolism in rats, used a diet that contained only 20 μg iron/g, no ascorbic acid, and 269 g vitamin-free casein/kg. Williams (1973), who found that vanadium deprivation did not affect iron metabolism or growth of the rat, fed a diet that contained 35 μg iron/g, 900 mg ascorbic acid/kg and 175 g vitamin-free casein/kg. Hopkins and Mohr (1971, 1974) used a diet that contained a high amount of methionine, and a high ratio of methionine to cystine (11.44 mg and 0.6 mg/g, respectively). The diet also contained a high amount of arginine (20 mg/g). In the study of Schwarz and Milne (1971), rats gained only 0.8 to 1.5 g/day, much less than the expected 3 to 5 g/day. The basal diet fed to the rats was not well described, but apparently was deficient in riboflavin (Moran and Schwarz, 1978). In the experiments done by Nielsen and co-workers (Nielsen and Ollerich, 1973; Nielsen and Sandstead, 1974; Myron et al., 1975: Nielsen and Myron, 1976; Nielsen and Uthus, 1977; Uthus et al., 1978; Nielsen et al., 1978; Nielsen, 1980), iron, methionine, arginine, cystine, copper, and ascorbic acid were all fed at variable and, perhaps, nonoptimal amounts.

The inability to determine whether the "deficiency signs" in early experiments with questionable diets were true deficiency signs, indirect changes caused by an enhanced need for vanadium in some metabolic function (Nielsen, 1988a), or manifestations of a pharmacologic action of vanadium made it difficult to conclude that vanadium is an essential nutrient (Nielsen, 1985a; Nechay et al., 1986). Based on recent findings described below, it is likely that all three phenomena were occurring in the studies.

The uncertainty about vanadium deficiency signs stimulated new efforts to produce vanadium deficiency signs in animals. These studies have used diets apparently containing adequate and balanced amounts of all known nutrients.

Anke and co-workers (1986) used goats in their vanadium deprivation studies. They found that, when compared to controls fed 2 μg vanadium/g diet, goats fed 10 ng vanadium/g diet exhibited a higher abortion rate and produced less milk during the first 56 days of lactation. More than 41% of kids from vanadium-deprived goats died between day 7 and 91 of life with some deaths preceded by convulsions; only 9% of kids from vanadium-supplemented goats died during this time. No significant changes in iron metabolism or serum cholesterol were seen. Serum creatinine and β-lipoprotein were elevated and serum glucose was depressed in the vanadium-deprived goats. Also, skeletal deformations were seen in the forelegs, and forefoot tarsal joints were thickened.

Uthus and Nielsen (1988, 1989) reported that, when compared to controls fed 1 μg vanadium/g diet, vanadium deprivation (2 ng vanadium/g diet) increased thyroid weight and thyroid weight/body weight ratio and tended to decrease growth. Dietary iodine was also varied in this study. As a result, some variables examined were affected by an interaction between vanadium and iodine. As dietary iodine increased from 50 ng to 0.33 μg to 25 μg/g diet, thyroid peroxidase activity decreased with the decrease more marked in the vanadium-supplemented (38.1 to 12.3 to 3.5 mGU/mg protein than the vanadium-deprived (18.7 to 10.2 to 6.8 mGU/mg protein) rats. Although not significant, both plasma thyroxine and triiodothyronine seemed to parallel thyroid peroxidase. Another finding was that, as dietary iodine increased, plasma glucose increased in the vanadium-deprived rats, but decreased in the vanadium-supplemented rats.

Unfortunately, the conclusiveness of the findings from the recent studies are somewhat diminished by the fact that the control diets, like those in the early vanadium essentiality studies, contained high amounts of vanadium (1 or 2 μg/g); these amounts may have had an unknown pharmacologic effect in these studies. Nonetheless, because the diets used appeared complete, and because several of the deprivation findings seem related (see below) and consistent, the recent vanadium deprivation studies probably have found some true deficiency signs. It is unlikely the diet lacked any nutrient that would cause such marked deficiency signs prevented by vanadium supplementation. The findings strongly suggest that vanadium is an essential nutrient for higher animals, including humans.

BIOCHEMICAL SIGNS

Because vanadium is such an active element in vitro and pharmacologically, there have been numerous biochemical and physiological functions suggested for it. Recently, a number of reviews (Boyd and Kustin, 1984; Nechay, 1984; Nielsen, 1984b, 1987, 1988b) have appeared discussing the evidence behind the suggestions that vanadium might have a role in the regulation of (Na, K)-ATPase, phosphoryl transfer enzymes, adenylate cyclase, and protein kinases. They also discuss the possible role of vanadium in hormone, glucose, lipid, bone and tooth metabolism, and as an enzyme cofactor in the form of vanadyl. However, no specific biochemical function has been identified for vanadium in higher animals. The recent discovery in lower forms of life of enzymes which require vanadium lends credence to the possibility that vanadium has a similar role in higher animals. These enzymes are nitrogenase in bacteria (Hales et al., 1986; Smith et al., 1987;

Eady, 1988) which reduces dinitrogen to ammonia, and iodoperoxidase and bromoperoxidase in algae (Vilter, 1984; De Boer *et al.*, 1986a, 1986b; Krenn *et al.*, 1987; Vilter and Rehder, 1987) and lichens (Plat *et al.*, 1987) which catalyze the oxidation of halide ions by hydrogen peroxide, thus facilitating the formation of a carbon-halogen bond.

The finding that some haloperoxidases require vanadium for activity suggests a similar function for vanadium in higher organisms. The best known haloperoxidase in animals is thyroid peroxidase. As described above, vanadium deprivation affected the response of thyroid peroxidase activity to changing dietary iodine in addition to elevating thyroid weight and thyroid weight/body weight ratio. Another finding which suggests that vanadium affects halogen metabolism in higher animals is that in pigeons, toxic amounts of vanadium increase the secretion of thyroid hormone and the activity of thyroid follicles (Diwan and Belsare, 1987).

The changes in bone in the vanadium-deprivation studies found by Anke and co-workers (1986) and in the early vanadium deprivation studies of Nielsen and Ollerich (1973) and Hunt (1979) may be related to changes in thyroid hormone metabolism. Thyroid hormones affect bone metabolism; this has been recently reviewed by Auwerk and Bouillon (1986). In humans, hypothyroidism can lead to a decrease in serum total and ionized calcium, serum phosphorus, and the urinary excretion of phosphorus and hydroxyproline. Hypothyroidism can also reduce both bone formation and resorption. In addition, thyroid hormones may directly enhance the production of somatomedins or other cartilage-regulating factors, or directly influence cartilage growth and maturation, which in turn affect bone formation.

Moreover, thyroid hormone also influences, or thyroid metabolism is affected by, many of the nutrients which interact with vanadium, including iron (Tang *et al.*, 1988) and sulfur amino acids (Yagasaki *et al.*, 1986; Suberville *et al.*, 1987). Also, many of the changes seen in early vanadium deprivation studies can be related to changes in thyroid metabolism, including changes in plasma cholesterol, and liver lipids and phospholipids (Alexander, 1984).

The suggestion that vanadium has a role in thyroid hormone metabolism needs further study because many of the actions of vanadium can be explained by it having a different role, e.g., in cell proliferation and differentiation similar to several growth factors. Vanadate mimics growth factors such as epidermal growth factor, fibroblast growth factor and insulin (Lau *et al.*, 1988; Nechay *et al.*, 1989). This is exemplified by the finding that orthovanadate stimulates bone cell proliferation and collagen synthesis in vitro (Canalis, 1985; Lau *et al.*, 1988). Another study (Kato *et al.*, 1987) showed that vanadate increased proteoglycan synthesis by stimulating the conversion of poorly differentiated chondrocyte cultures from a "fibroblastic" expression to a "chondrocyte" expression. In this study, vanadyl was as potent as vanadate, both being present in physiological amounts, in stimulating ^{35}S-sulfate incorporation into proteoglycans in rabbit chondrocyte cultures. Low concentrations, similar to those found in human serum, were found to be required for optimal growth of fibroblasts in tissue culture (McKeehan *et al.*,

1977). All of the above observations suggest that vanadium has a physiological role as a growth factor or is necessary for expression of a growth factor.

2. Metabolism

ABSORPTION

Limited information exists about the metabolism of physiological amounts of vanadium in animals. Nonetheless, it is apparent that most ingested vanadium is unabsorbed and is excreted via the feces (Nielsen, 1987, 1988b). Based on the very low concentrations of vanadium normally in urine, in comparison with the estimated daily intake and fecal levels of vanadium, apparently less than 5% of vanadium ingested is absorbed. Byrne and Kosta (1978) estimated that no more than 1% of vanadium normally ingested with the diet is absorbed. Curran et al. (1959) reported that about 0.1 to 1.0% of vanadium in 100 mg of the very soluble diammonium oxytartarovanadate was absorbed by the human gastrointestinal tract. The human studies agree well with some animal studies. Hansard et al. (1982) estimated that only 0.13% to 0.75% (mean of 0.34%) of ingested vanadium as ammonium metavanadate was absorbed from the sheep gut. Conklin et al. (1982) found that the uptake of radioactive V_2O_5 administered orally to rats was 2.6% of the dose. Similar results were obtained by other investigators (Parker and Sharma, 1978; Roschin et al., 1980). However, two other studies with rats indicated that a much greater amount of vanadium can be absorbed from the gastrointestinal tract. Wiegmann et al. (1982) fasted rats overnight and then gavaged them with 5 μmol of Na_3VO_4 in 1.0 ml 0.9% NaCl containing 1.0 μCi [48]V. After 4 days, 86.6% $\pm$ 2% of the administered [48]V was recovered in the feces and urine. Only 69.1% $\pm$ 1.8% of the dose was recovered in the feces; recovery from feces increased to 85.7% $\pm$ 1.5% if 1.0 ml of a suspension of $Al(OH)_3$ was administered simultaneously with the vanadium. In either case, the findings indicated an absorption of greater than 10%. Bodgen et al. (1982) found that rats retained 39.7% $\pm$ 18.5% and excreted in the feces 59.1% $\pm$ 18.8% of vanadium ingested as sodium metavanadate supplemented at 5 or 25 μg/g of a casein-sucrose-dextrin based diet. The finding of high absorption of vanadium was also reported by Proescher et al. (1917). They found that 12.4% of an orally administered dose of sodium metavanadate (12.5 mg/day for 12 days) to a healthy young man was recovered in urine. In the animal studies described, factors such as fasting, dietary composition, and chemical form of ingested vanadium probably affected the percentage of ingested vanadium absorbed from the intestine. Regardless, the rat studies and the early work of Proescher et al. (1917) suggest caution in assuming that ingested vanadium always will be poorly absorbed from the gastrointestinal tract. Kinetic modeling of whole body vanadium metabolism in sheep indicated that a significant amount of vanadium absorption occurs in the upper gastrointestinal tract (Patterson et al., 1986).

TRANSPORT

Dietary vanadium probably occurs mainly as VO^{2+} (vanadyl, V^{4+}) or as HVO_4^{2-} (vanadate, V^{5+}). Most ingested vanadium probably is transformed in the stomach to VO^{2+} and remains in this form as it passes into the duodenum (Chasteen et al., 1986). However, V^{5+} is absorbed 3 to 5 times more effectively than V^{4+}. Thus, the effect of other dietary components on the form of vanadium in the stomach, and speed at which it is transformed into V^{4+}, probably markedly affect the percentage of ingested vanadium absorbed (Chasteen et al., 1986). Evidence supporting the preceding are the findings showing that a number of substances can ameliorate vanadium toxicity, including ascorbic acid (Hill, 1979b), EDTA (Hathcock et al., 1964), chromium (Hill, 1979a), protein (Hill, 1979c), ferrous iron (Nielsen et al., 1983; Nielsen, 1985; Blalock and Hill, 1987), chloride (Hill, 1985), and aluminum hydroxide (Wiegmann et al., 1982).

Much evidence suggests that the binding of the vanadyl ion to iron-containing proteins is important in vanadium metabolism (Sabbioni and Marafante, 1978, 1981; Harris et al., 1984; Chasteen et al., 1986). Animal experiments indicate that about 90% of the vanadium in blood is associated with the plasma fraction (Hathcock et al., 1964; Chasteen et al., 1986). Vanadium probably exists in both oxidation states in blood as a result of oxygen tension and the presence of reducing agents such as ascorbate and catecholamines. Regardless of the oxidation state entering the blood, vanadium apparently is converted into vanadyl-transferrin and vanadyl-ferritin complexes in plasma and body fluids (Sabbioni and Marafante, 1978, 1981; Harris et al., 1984; Chasteen et al., 1986). About 40% to 50% of vanadium is associated with transferrin. If vanadate appears in the blood, it probably is quickly converted into vanadyl, most likely in the erythrocytes (Harris et al., 1984). It remains to be determined whether ferritin is a storage vehicle for vanadium as well as for iron in the liver and whether vanadyl-transferrin can transfer vanadium through the transferrin receptor.

RETENTION

A review of recent analyses using reliable techniques indicates that very little vanadium is retained under normal conditions in the body; most tissues contain less than 10 ng vanadium/g wet weight. For example, analyses by the sensitive techniques of neutron activation (Byrne and Kosta, 1978; Vanoeteren et al., 1982) have found the following concentrations of vanadium in human tissues (ng/g wet weight): fat and muscle, 0.55; heart, 1.1; kidney, 3; liver, 7.5; lung, 2.1; and thyroid, 3.1. Reported mean concentrations of vanadium in scalp hair of healthy adults range from 433 pg/g to 90 ng/g (Moo and Pillay, 1983; Marumo et al., 1984; Jervis et al., 1985; Ward et al., 1985). Neutron activation techniques were used to show that human colostrum, and transitional and mature milk, generally contained less than 1.0 ng vanadium/g dry weight (Kosta et al., 1983), and that human serum vanadium is in the range of 0.016 to 0.939 ng/ml, with most values below 0.15 ng/ml (Cornelis et al., 1981).

Occupational exposure or pathological conditions can alter the vanadium concentration in blood and organs of humans. When compared to healthy or normal controls, the vanadium concentration was higher in the hair of welders (Jervis et al., 1985) and lower in the hair of multiple sclerosis patients (Ward et al., 1985). Marumo et al. (1984) found that the vanadium concentration in hair was higher in nondialyzed and hemodialyzed patients with chronic renal failure than in healthy controls or hemofiltered patients with chronic renal failure.

Animal studies indicate that tissue vanadium is markedly elevated in animals fed high dietary vanadium. In rats, liver vanadium increased from 10 to 55 ng/g wet weight when dietary vanadium was increased from 0.1 to 25 $\mu g/g$ (Bodgen et al., 1982). In sheep, bone vanadium increased from 220 to 3320 ng/g dry weight when dietary vanadium was increased from 10 to 220 $\mu g/g$ (Hansard et al., 1978). Bone apparently is a major sink for retained vanadium (Hansard et al., 1978; Parker and Sharma, 1978; Hansard et al., 1982; Wiegmann et al., 1982; Edel et al., 1984). Rat tissue vanadium is affected also by age. In rats between ages 21 and 115 days the vanadium concentration decreased in kidney, liver, lung and spleen, and increased in fat and bone (Edel et al., 1984).

EXCRETION

Most ingested vanadium is excreted via the feces; the feces probably contain mostly unabsorbed vanadium. Based upon studies in which vanadium is administered parenterally, urine is the major excretory route for absorbed vanadium. At 96 hours after an intravenous injection of ^{48}V in rats, 30 to 46% of the dose had been excreted in the urine and 9 to 10% in the feces (Hopkins and Tilton, 1966). Vanadium administered subcutaneously, intramuscularly, intraperitoneally, or intratracheally is generally metabolized as when administered intravenously (Pepin et al., 1977; Roschin et al., 1980; Wiegmann et al., 1982). For example, within 5 days after intraperitoneal administration of ^{48}V-labeled vanadate to rats, 41% of the dose appeared in the urine and 8.3% in the feces (Wiegmann et al., 1982). Another study showed that 66% of ^{48}V intramuscularly injected as $VOCl_2$ was eliminated in the urine within 24 hours (Roschin et al., 1980). Both high and low molecular weight complexes have been found in urine (Sabbioni and Marafante, 1978; Chasteen et al., 1986); one of these complexes may be transferrin. Chasteen et al. (1986) found the majority of vanadium was excreted as a low molecular weight VO^{2+} complex. The fecal excretion of injected vanadium probably is that excreted through the bile.

DIETARY INTAKE AND REQUIREMENTS

Any human requirement for vanadium would likely be very small. The diets used in animal deprivation studies contained only 2 to 25 ng vanadium/g; these often did not markedly affect the animals. Vanadium deficiency has not been identified in humans; yet, as indicated below, most diets supply less than 30 μg daily, with most near 15 μg daily. The preceding observations suggest that a dietary intake of 10 μg daily probably meets any postulated vanadium requirement.

The daily intake of vanadium is relatively low in comparison with other essential trace elements. Byrne and Kosta (1978) stated that their analyses indicated that the daily dietary intake of vanadium is in the order of "a few tens of micrograms and may vary widely". Myron et al. (1978) found that nine institutional diets supplied 12.4 to 30.1 μg of vanadium daily, and intake averaged 20 μg. Ten diets of the United Kingdom Total Diet Study were found to supply an average of 13 μg vanadium daily (Evans et al., 1985). Estimating the daily intakes of vanadium for eight age-sex groups resulted in a range of 6.2 to 18.3 μg in the U.S. Food and Drug Administration's Total Diet Study (Pennington and Jones, 1987). Myron et al. (1978) found that beverages, fats and oils, and fresh fruits and vegetables contained the least vanadium, ranging from < 1.0 to 5 ng/g. Whole grains, seafood, meats and dairy products were generally within a range of 5 to 30 ng/g. Byrne and Kosta (1978) obtained similar results. They found that most fats, oils, fruits, and vegetables contained < 1.0 ng/g. Cereals, liver, and fish tended to have intermediate levels of about 5 to 40 ng/g. Only a few items, such as spinach, parsley, mushrooms and oysters contained relatively high amounts of vanadium.

3. Summary

Recent findings indicate that vanadium is an essential nutrient. Vanadium deficiency signs apparently include impaired reproduction characterized by higher abortions and perinatal mortality, bone abnormalities, and changes in thyroid metabolism. The deprivation signs suggest that vanadium has a biological function that affects thyroid hormone metabolism or acts as a growth factor in a manner similar to fibroblast growth factor.

Vanadium absorption apparently is less than 5% from the gastrointestinal tract; however, some studies indicate it could be higher. Vanadium is metabolized as VO^{2+} and HVO_4^{2-} in vivo, utilizing the iron transport and storage proteins transferrin and ferritin. Very little vanadium is retained in the body; bone apparently is a major sink for retained vanadium. Excretion of absorbed vanadium occurs mainly through the urine as small molecular weight complexes.

The dietary requirement for vanadium is very small, probably near 10 μg daily. Most diets supply less than 30 μg daily, with most near 15 μg daily. Because this is near the postulated requirement, further studies are warranted to ascertain whether vanadium is of practical nutritional concern.

References

Alexander, N. M.: 1984, *Biochemistry of the Essential Ultratrace Elements*, E. Frieden, ed, Plenum, New York, pp 33–53.

Anke, M.; Groppel, B.; Gruhn, K.; Kosla, T.; Szilagyi, M.: 1986, *Spurenelement-Symposium: New Trace Elements*. Anke, M.; Baumann, W.; Braunlich, H.; Bruckner, C.; Groppel, B.; eds. Jena, Friedrich-Schiller-Universitat, pp 1266–1275.

Auwerx, J.; Bouillon, R.: 1986, *Quart. J. Med.* 60, 737–752.
Bertrand, D.: 1950, *Bull. Amer. Mus. Nat. Hist.* 94, 403–456.
Blalock, T. L.; Hill, C. H.: 1987, *Biol. Trace Element Res.* 14, 225–235.
Bodgen, J. D.; Higashino, H.; Lavenhar, M. A.; Bauman, J. W., Jr.; Kemp, F. W.; Aviv, A.: 1982, *J. Nutr.* 112, 2279–2285.
Boyd, D. W.; Kustin, K., 1984, *Adv. Inorg. Biochem.* 6, 311–365.
Byrne, A. R.; Kosta, L.: 1978, *Sci. Total Environ.* 10, 17–30.
Canalis, E.: 1985, *Endocrinology* 116, 855–862.
Chasteen, N. D.; Lord, E. M.; Thompson, H. J.: 1986, *Frontiers in Bioinorganic Chemistry*, A. V. Xavier, ed, VCH Verlagsgesellschaft, Weinhein FRG, pp 133–141.
Conklin, A. W.; Skinner, S. C.; Felten, T. L.; Sanders, C. L.: 982, *Toxicol. Lett.* 11, 199–203.
Cornelis, R.; Versieck, J.; Mees, L.; Hoste, J.; Barbier, F.: 1981, *Biol. Trace Element Res.* 3, 257–263.
Curran, G. L.; Azarnoff, D. L.; Bolinger, R. E.: 1959, *J. Clin. Invest.* 38, 1251–1261.
De Boer, E.; Tromp, M. G. M.; Plat, H.; Krenn, G. E.; Wever, R.: 1986a, *Biochem. Biophys. Acta* 872, 104–115.
De Boer, E.; Van Kooyk, Y.; Tromp, M. G. M.; Plat, H.; Wever, R.: 1986b, *Biochem. Biophys. Acta* 869, 48–53.
Diwan, M.; Belsare, D. K.: 1987, *J. Environ. Biol.* 8, 157–166.
Eady, R. R.: 1988, *Biofactors* 1, 111–116.
Edel, J.; Pietra, R.; Sabbioni, E.; Marafante, E. Springer A.; Ubertalli, L.: 1984, *Chemosphere* 13, 87–93.
Evans, W. W.; Read, J. I.; Caughlin, D.: 1985, *Analyst* 110, 873–877.
Hales, B. J.; Case, E. E.; Morningstar, J. E.; Dzeda, M. F.; Mauterer, L. A.: 1986, *Biochemistry* 25, 7251–7255.
Hansard, S. L., II; Ammerman, C. B.; Fick, K. R.; Miller, S. M.: 1978, *J. Anim. Sci.* 46, 1091–1095.
Hansard, S. L., II; Ammerman, C. B.; Henry, P. R.: 1982, *J. Anim. Sci.* 55, 350–356.
Harris, W. R.; Friedman, S. B.; Silberman, D.: 1984, *J. Inorg. Biochem.* 20, 157–169.
Hathcock, J. N.; Hill, C. H.; Matrone, G.: 1964, *J. of Nutr.* 82, 106–110.
Heyliger, C. E.; Tahiliani, A. G.; McNeill, J. H.: 1985, *Science* 227, 1474–1477.
Hill, C. H.: 1979a, *Chromium in Nutrition and Metabolism*, D. Shapcott, J. Hubert, eds, Elsevier/North-Holland Biomedical Press, Amsterdam, pp 229–240.
Hill, C. H.: 1979b, *J. Nutr.* 109, 84–90.
Hill, C. H.: 1979c, *J. Nutr.* 109, 501–507.
Hill, C. H.: 1985, *Nutr. Res., Suppl. I*, 555–559.
Hill, C. H.: 1988, *Trace Elements in Man and Animals 6*, L. S. Hurley, C. L. Keen, B. Lonnerdal, R. B. Rucker, eds, Plenum New York, pp 585–587.
Hopkins, L. L., Jr.; Mohr, H. E.: 1971, *Newer Trace Elements in Nutrition*, W. Mertz, W. E. Cornatzer, eds, Marcel Dekker, New York, pp 195–213.
Hopkins, L.L., Jr.; Mohr, H. E.: 1974, *Fed. Proc.* 33, 1773–1775.
Hopkins, L. L., Jr.; Tilton, B. E.: 1966, *Amer. J. Physiol.* 211, 169–172.
Hunt, C. D.: 1979, *The Effect of Dietary Vanadium on* ^{48}V *Metabolism and Proximal Tibial Growth Plate Morphology in the Chick*, Ph.D. Thesis, University of North Dakota, Grand Forks, ND.
Jervis, R. E.; Evans, G. J.; Hewitt, P. J.: 1985, *Nutrition Res. Suppl. I*, 627–633.
Kato, Y.; Iwamoto, M.; Koike, T.; Suzuki, F.: 1987, *J. Cell Biol.* 104, 311–319.
Kosta, L.; Byrne, A. R.; Dermelj, M.: 1983, *Sci. Total Envir.* 29, 261–268.
Krenn, B. E.; Plat, H.; Wever, R.: 1987, *Biochem. Biophys. Acta* 912, 287–291.
Lau, K. H. W.; Tanimoto, H.; Baylink, D. J.: 1988, *Endocrinology* 123, 2858–2867.
Marumo, F.; Tsulsamoto, Y.; Iwanami, S.; Kishimoto, T.; Yamagami, S.: 1984, *Nepron* 38, 267–272.
McKeehan, W. L.; McKeehan, K. A.; Hammond, S. L.; Ham, R. G.: 1977, *In Vitro* 3, 399–416.
Moo, S. P.; Pillay, K. K. S.: 1983, *J. Radioanal. Chem.* 77, 141–147.
Moran, J. K.; Schwarz, K.: 1978, *Fed. Proc.* 37, 671 abs.
Myron, D. R.; Givand, S. H.; Hopkins, L. L.; Nielsen, F. H.: 1975, *Fed. Proc.* 34, 923 abs.
Myron, D. R.; Zimmerman, T. J.; Shuler, T. R.; Klevay, L. M., Lee, D. E., Nielsen, F. H.: 1978, *Am. J. Clin. Nutr.* 31, 527–531.

Nechay, B. R.: 1984, *Ann. Rev. Pharmacol. Toxicol.* 24, 501–524.

Nechay, B. R.; Nanninga, L. B.; Nechay, P. S. E.; Post, R. L.; Grantham, J. J.; Macara, I. G.; Kubena, L. F.: Phillips, T. D.; Nielsen, F. H.: 1986, *Fed. Proc.* 45, 123–132.

Nechay, B. R.; Norcross-Nechay, K.; Nechay, P. S. E.: 1989, *Spurenelement-Symposium: Molybdenum, Vanadium, and Other Trace Elements,* Friedrich-Schiller-Universitat, Jena DDR, in press.

Nielsen, F. H.: 1980, *Advances in Nutritional Research,* Vol. 3, H. H. Draper, ed, Plenum, New York, pp 157–172.

Nielsen, F. H.: 1984a, *Proc. ND Acad. Sci.* 38, 57.

Nielsen, F. H.: 1984b, *Ann. Rev. Nutr.* 4, 21–41.

Nielsen, F. H.: 1985a, *J. Nutr.* 115, 1239–1247.

Nielsen, F. H.: 1985b, *Nutr. Res. Suppl. I,* M. Abdulla, B. M. Nair, R. K. Chandra, eds, Pergamon Press, New York, pp 527–530.

Nielsen, F. H.: 1987, *Trace Elements in Human and Animal Nutrition,* Ed. 5, W. Mertz, ed., Academic Press, San Diego, pp 275–300.

Nielsen, F. H.: 1988a, *Nutr. Rev.* 46, 327–341.

Nielsen, F. H.: 1988b, *Trace Minerals in Foods,* K. T. Smith, ed. Marcel Dekker, New York, pp 357–428.

Nielsen, F. H.; Myron, D. R.: 1976, *Fed. Proc.* 35, 683 abs.

Nielsen, F. H.; Myron, D. R.; Uthus, E. O.: 1978, *Trace Element Metabolism in Man and Animals-3,* M. Kirchgessner, ed, Technical University of Munchen, Freising,-Weihenstephan, BRD, pp 244–247.

Nielsen, F. H.; Ollerich, D. A.: 1973, *Fed. Proc.* 32, 929 abs.

Nielsen, F. H.; Sandstead, H. H.: 1974, *Amer. J. Clin. Nutr.,* 27, 515–520.

Nielsen, F. H.; Uhrich, K. E.; Shuler, T. R.; Uthus, E. O.: 1983, *Spurenelement Symposium 1983,* M. Anke, W. Baumann, H. Braunlich, Chr. Bruckner, eds, Friedrich-Schiller-Universitat, Jena, DDR, pp 127–134.

Nielsen, F. H.; Uhrich, K.: Uthus, E. O.: 1984, *Biol. Trace Elements Res.* 6, 117–132.

Nielsen, F. H.; Uthus, E. O.: 1977, *Fed. Proc.* 36, 1123 abs.

Parker, R. D. R.; Sharma, R. P.: 1978, *J. Environ. Pathol. Toxicol.* 2, 235–245.

Patterson, B. W.; Hansard, S. L., II; Ammerman, C. B.; Henry, P. R.; Zech, L. A.; Fisher, W. R., 1986, *Amer. J. Physiol.* 251, R325–R332.

Pennington, J. A. T.; Jones, J. W., 1987, *J. Amer. Diet. Assn.* 87, 1644–1650.

Pepin, G.; Bouley, G.; Boudene, C.: 1977, *C. R. Hebd. Seances Acad. Sci.,* Paris 285, 451–454.

Plat, H.; Krenn, B. E.; Wever, R.: 1987, *Biochem. J.* 248, 277–279.

Proescher, F.; Seil, H. A.; Stillians, A. W.: 1917, *Am. J. Syph.* 1, 347–405.

Roschin, A. V.; Ordzhonikidze, E. K.; Shalganova, I. V.: 1980, *J. Hyg. Epidemiol. Microbiol. Immunol.* 24, 377–383.

Rygh, O.: 1949, *Bull. Ste'. Chem. Biol.* 31, 1403–1407.

Sabbioni, E.; Marafante, E.: 1978, *Bioinorg. Chem.* 9, 389–407.

Sabbioni, E.; Marafante, E.: 1981, *J. Toxicol. Environ. Health* 8, 419–429.

Schroeder, H. A.; Balassa, J. J.; Tipton, I. H.: 1963, *J. Chron. Dis.* 16, 1047–1071.

Schwarz, K.; Milne, D. B.: 1971, *Science* 174, 426–428.

Shuler, T. R.; Nielsen, F. H.: 1983 *Proc. ND Acad. Sci.* 37, 88.

Shuler, K.; Nielsen, F. H.: 1985, *Trace Elements in Man and Animals-5,* C. F. Mills, I. Bremner, J. K. Chesters, eds, Commonwealth Agricultural Bureau, United Kingdom, pp 382–384.

Smith, B. E.; Campbell, F.; Eady, R. R.; Eldridge, M.; Ford, C. M.; Hill, S.; Kavanagh, E. P.; Lowe, D. J.; Miller, R. W.; Richardson, T. H.; Robson, R. L.; Thorneley, R. N. F.; Yates, M. G.: 1987, *Phil. Trans. R. Soc. Lond. B.* 317, 131–146.

Strasia, C. A.: 1971, *Vanadium: Essentiality and Toxicity in the Laboratory Rat,* Ph.D. Thesis, Purdue University, Lafayette, IN.

Suberville, C.; Higueret, P.; Taruoura, D.; Garan, H.: 1987, *Brit. J. Nutr.* 58, 105–111.

Tang, F.; Wong, T. M.; Loh, T. T.: 1988, *Horm. metabol. Res.* 20, 616–619.

Uthus, E. O.; Nielsen, F. H.: 1988, *FASEB J.* 2, A841.

Uthus, E. O.; Nielsen, F. H.: 1989, 6. *Spurenelement-Symposium: Molybdenum, Vanadium, and Other Trace Elements*, Friedrich- Schiller-Universitat, Jena, DDR, in press.

Uthus, E. O.; Nielsen, F. H.; Myron, D. R.: 1978, *Fed. Proc.* 37, 893 abs.

Vanoeteren, C.; Cornelis, R.; Versieck, J.; Hoste, J.; De Roose, J.: 1982, *Radioanal. Chem.* 70, 219–238.

Vilter, H.: 1984, *Phytochem.* 23, 1387–1390.

Vilter, H.; Rehder, D.: 1987, *Inorg. Chim. Acta* 136, L7–L10.

Ward, N. I.; Bryce-Smith, O.; Minski, M.; Matthews, W. B.: 1985, *Biol. Trace Element Res.* 7, 153–159.

Wiegmann, T. B.; Day, H. D.; Patak, R. V.: 1982, *J. Toxicol. Environ. Health* 10, 233–245.

Williams, D. L.: 1973, *Biological Value of Vanadium for Rats, Chickens, and Sheep*, Ph.D. Thesis, Purdue University, Lafayette, IN.

Yagasaki, K.; Ohsawa, N.; Funabiki, R.: 1986, *Nutr. Reports International* 33, 321–328.

IV. Vanadates as Phosphate Analogs in Biochemistry

MICHAEL J. GRESSER
Merck Frosst Centre for Therapeutic Research, P.O. Box 1005, Pointe Claire-Dorval, Quebec, Canada H9R 4P8

and

ALAN S. TRACEY
Dept. of Chemistry, Simon Fraser University, Burnaby, B.C., Canada V5A 1S6

Introduction

There are two general kinds of ways in which vanadates can serve as phosphate analogs in biochemical systems. In the case of enzymes or other proteins which do not catalyse cleavage of a bond to the phosphorus atom, and for which phosphates are the natural ligands, vanadates can serve as alternate ligands or substrates. Inorganic vanadate (V_i) itself can function as an analog of inorganic phosphate (P_i) in these systems, and spontaneously formed vanadate complexes can act as analogs of phosphate esters. The vanadate can in some systems be as "good" a ligand as the corresponding phosphate.

In the case of enzymes which catalyse phosphoryl transfer reactions, V_i and its complexes can function as inhibitors. Extremely low concentrations of vanadate complexes are sufficient to strongly inhibit some phosphoryl transfer enzymes, and in these cases the bound vanadate complex is considered to be a transition-state analog.

In this chapter examples of both of these types of behaviour will be considered, as well as cases in which such behaviour could potentially occur but has not been observed. The utility of vanadates as mechanistic probes will be discussed, and consideration will be given to the possible physiological significance of the behaviour of vanadates as phosphate analogs. In addition to the other chapters in this volume, the interested reader is referred to a number of useful reviews on the chemistry and biochemistry of vanadium (Pope and Dale 1968; Kepart 1973; Baes and Mesmer 1976; Ramasarma and Crane 1981; Chasteen 1981; Rehder 1982; Chasteen 1983; Kustin *et al.* 1983; Jandhyala and Hom 1983; Erdmann *et al.* 1984; Nechay 1984; Nechay *et al.* 1986).

N. Dennis Chasteen (ed.), Vanadium in Biological Systems, 63–79.
© 1990 *Kluwer Academic Publishers. Printed in the Netherlands.*

Vanadates as Substrate Analogs

VANADATE AND ARSENATE AS SUBSTITUTES FOR INORGANIC PHOSPHATE

It has been reported that V_i is a substrate, replacing P_i, in the reactions catalysed by glyceraldehyde-3-phosphate dehydrogenase (DeMaster and Mitchell 1973), diadenosine tetraphosphate phosphorylase (Guranowski and Blanquet 1986), and nucleoside phosphorylase (Drueckhammer *et al.* 1989). Similar behaviour has been observed when arsenate (As_i) was used in these systems and in others (Doudoroff *et al.* 1947; Slocum and Varner 1960; Itada and Cohn 1963). It is reasonable that V_i should behave similarly to P_i in systems in which P_i normally acts as a nucleophile, because V-O bond lengths in vanadates are not much longer than P-O bond lengths in phosphate, and the pKa values for phosphoric and vanadic acids are similar, vanadic acid having somewhat higher pKa values (Chasteen 1983). The products of the nucleophilic attack of As_i or V_i on the reactive intermediates in these systems behave differently from the phosphorylated products formed in the normal reactions. The arsenate and vanadate analogs of the phosphorylated compounds hydrolyse rapidly relative to the phosphates. This behaviour had been predicted in the case of arsenates (Warburg and Christian 1939), and in the case of vanadates is consistent with the known fast rate of water exchange on vanadium(V) (Kustin *et al.* 1983).

Arsenate esters hydrolyse with rate constants about 10^5-fold larger than those for hydrolysis of phosphate esters (Long and Ray 1973; Lagunas 1980; Langunas *et al.* 1984), and a similar relationship appears to hold for anhydrides (Moore *et al.* 1983; Slooten and Nuyten 1983). The rate constants for hydrolysis of vanadate esters (Nour-Eldeen *et al.* 1985; Gresser and Tracey 1985) and anhydrides (Gresser *et al.* 1986) appear to be about 10^{10}-fold larger than those for the corresponding phosphate compounds. The reasons for these very large differences in rates of ligand exchange reactions of phosphate, arsenate, and vanadate are not clear, but the differences in the range of electronic energy levels of the orbitals of the central atom which are involved in the ligand exchange reaction might be an important factor. These are the 3s, 3p, and 3d orbitals of phosphorus, the 4s, 4p, and 4d orbitals of arsenic, and the 3d, 4s, and 4p orbitals of vanadium. The electron pair of the incoming ligand must be accommodated by an orbital of the central atom. When the unoccupied orbital of lowest energy level on the central atom is close in energy level to the orbitals occupied by the valence electrons, it is reasonable that ligand exchange reactions should be faster than when the lowest unoccupied orbital is at a much higher energy level than the valence electrons.

This fast spontaneous hydrolysis of the products of enzymic reactions in which As_i or V_i acts as a P_i analog regenerates the As_i or V_i, with the result that these P_i analogs participate catalytically rather than stoichiometrically in the reaction. Another consequence of the fast hydrolysis of the products of these reactions is that the products do not accumulate. This eliminates product inhibition as well as the slowing or stopping of net reaction due to the law of mass action in cases where the equilibrium constant for the reaction involving P_i is not large, such as the glyceraldehyde-3-phosphate dehydrogenase-catalysed reaction.

The behaviour of As_i and V_i as substitutes for P_i in enzymic reactions in which P_i acts as an electrophile; with cleavage of a phosphorus-oxygen bond, has been less thoroughly studied. Arsenate has been found to substitute for P_i in mitochondrial oxidative phosphorylation (Moore et al. 1983; Gresser 1981) and in photophosphorylation in Rhodospirillum Rubrum chromatophores (Slooten and Nuyten 1983; Slooten and Nuyten 1984). The resulting ADP-arsenate undergoes rapid spontaneous hydrolysis, which accounts for the uncoupling of mitochondrial oxidative phosphorylation by As_i. The ADP-arsenate can act as a substrate for hexokinase (Moore et al. 1983; Gresser 1981) or luciferase (Slooten and Nuyten 1983; Slooten and Nuyten 1984) and this provides the possibility to quantitate its formation. Attempts to detect similar behaviour by V_i in mitochondrial oxidative phosphorylation were not successful. It was found, however, that V_i bound to mitochondrial F_1-ATPase in competition with P_i, and that V_i inhibited F_1-catalysed ATP hydrolysis and ATP synthesis and hydrolysis by submitochondrial particles (Bramhall 1987). It thus appears that V_i binds at the catalytic site of the mitochondrial ATP synthase, but no ADP-vanadate could be detected by hexokinase trapping (Bramhall 1987), and there are conflicting reports regarding the ability of V_i to uncouple mitochondrial oxidative phosphorylation (DeMaster and Mitchell 1973; Hathcock et al. 1966). It is possible that ADP-vanadate is formed by submitochondrial particles, but that the resulting ADP-vanadate hydrolyses too rapidly to be detected. The interaction of vanadate complexes with ATP binding sites of enzymes is considered further in later sections of this chapter.

VANADATE AND ARSENATE ESTERS AND ANHYDRIDES AS SUBSTITUTES FOR THE CORRESPONDING PHOSPHATES

The first report of a vanadate ester which acted as a substrate for an enzyme whose natural substrate is the corresponding phosphate ester resulted from an attempt to determine whether V_i could substitute for P_i in mitochondrial oxidative phosphorylation (Nour-Eldeen et al. 1985). The coupling enzymes hexokinase and glucose 6-phosphate dehydrogenase (G6PDH) were present, and initially a positive result was obtained. When control experiments were done, it was found that the presence of glucose and G6PDH were sufficient for V_i-dependent NADP$^+$ reduction to occur. The observations were interpreted in terms of spontaneous condensation of V_i and glucose to form glucose 6-vanadate, which is a substrate for G6PDH. At high G6PDH concentrations, formation of glucose 6-vanadate became partially rate-limiting, and the rate constant for the vanadate ester formation reaction could be obtained by extrapolating to infinite enzyme concentration. It was found that the rate constant was about 10^{10}-fold larger than that for spontaneous formation of glucose 6-phosphate. Vanadate did not activate the G6PDH-catalysed oxidation of 6-deoxyglucose, consistent with the formation of glucose 6-vanadate in the vanadate-activated oxidation of glucose. Similar results were obtained from studies using glycerol 3-phosphate dehydrogenase (Craig 1986). In the case of this enzyme it was found that V_i activated the glycerol

3-phosphate dehydrogenase-catalysed oxidation of racemic 1,2-propanediol, but not of (S)-(+)-1,2-propanediol (Craig and Gresser unpub.). Presumably the substrate in the experiment with the racemic 1,2-propanediol was the vanadate ester of the (R)-(-)-1,2-propanediol, which is analogous to 1-deoxyglycerol 3-phosphate. This experiment with (S)-(+)-1,2-propanediol was analogous to the use of 6-deoxyglucose with G6PDH, and similar results were obtained.

These studies were confirmed, and extended to a number of other enzymes by D. Crans and coworkers (Drueckhammer et al. 1989). Additional systems in which this group has obtained evidence that spontaneously formed vanadate esters act as enzyme substrates include phosphoglucose isomerase, phosphoribose isomerase, and ribulose-5-phosphate epimerase. Studies with adenylate kinase yielded results consistent with adenosine-5-vanadate being a substrate for this enzyme (Craig and Gresser unpub.). Thus vanadate plus adenosine activate adenylate kinase as an ATPase. Also, glucose 6-vanadate can apparently bind to and be phosphorylated by the phospho form of phosphoglucomutase (Percival et al. 1989 sub.). It had been observed earlier that vanadate strongly activated the transfer of phosphate from the phospho form of phosphoglucomutase to glucose (Layne and Najjar 1979), but the results were not interpreted in terms of glucose 6-vanadate substituting for glucose 6-phosphate as a substrate. This reaction proceeds for only one turnover, because after phosphorylation a change in the mode in which the resulting glucose 1-phosphate 6-vanadate is bound occurs, such that the vanadate moiety is at a phosphoryl transfer site and acts as a transition state analog inhibitor.

The spontaneous formation of arsenate esters which appear to act as substrates for enzymes has been observed in studies of phosphoglucomutase (Long and Ray 1973), glucose 6-phosphate dehydrogenase, 6-phosphogluconate dehydrogenase (Lagunas 1980), adenylate kinase, adenylate deaminase (Lagunas et al. 1984), 3-phosphoglycerate dehydrogenase, phosphoglucose isomerase, phosphoribose isomerase, ribulose 5-phosphate 3-epimerase, fuculose 1-phosphate aldolase, and rhamnulose 1-phosphate aldolase (Drueckhammer et al. 1989). Studies by Lagunas and coworkers had provided earlier indications of this kind of behaviour in some systems (Lagunas et al. 1984, and references therein). It thus appears that in general, when arsenate or vanadate esters can form spontaneously they will, and they will be recognized as substrates by enzymes whose normal substrates are the corresponding phosphate esters. This behaviour has been applied to the use of enzymes in organic synthesis (Drueckhammer et al. 1989), although the fact that vanadate is an oxidizing agent limits its usefulness in such systems. Arsonates, like arsenates, also undergo rapid transesterification reactions, and this has been exploited in chemical (Jacobson et al. 1979a; Jacobson et al. 1979b) and enzymatic systems (Rozovskaya et al. 1984; Adams et al. 1984; Visedo-Gonzalez and Dixon 1989).

Arsenate or vanadate anhydrides are presumably formed as products, which hydrolyse rapidly and spontaneously in some of the systems mentioned above, such as the activation of an ATPase activity of adenylate kinase by adenosine and vanadate or arsenate. However, spontaneously formed vanadate or arsenate anhydrides have not yet been observed to act as substrates for enzymes, although the ADP-arsenate formed in oxidative phosphorylation and photophosphorylation can act as a substrate for hexokinase or luciferase.

[51]V NMR Studies of Vanadate Analogs of Phosphates

Although a number of physical probes have been used in studies of vanadates, [51]V nmr has been the most revealing. The use of this probe was reviewed in 1982 (Rehder 1982), and again in this volume. In this section the only aspect of the [51]V nmr studies considered will be their application in studies of vanadate complexes which can usefully be thought of as analogs of phosphate esters and anhydrides. Vanadate esters formed by allowing stoichiometric amounts of alcohols to react with $VOCl_3$ in nonaqueous solvents were studied by Rehder (Rehder 1982; Rehder 1977), but the first report of spontaneously formed vanadate esters in aqueous solution was the result of a study of V_i in aqueous ethanol (Gresser and Tracey 1985). It was found that addition of ethanol to a solution of aqueous vanadate resulted in the appearance of two new [51]V signals. The ratio of the intensity of one of the new signals to that of the signal due to V_i increased linearly with the ethanol/water concentration ratio. The ratio of the intensity of the other new signal to that of the V_i signal increased linearly with the second power of the ethanol/water concentration ratio. The slopes of the correlations were not affected by changing the vanadium atom concentration in the solution. These results are consistent with the formation of ethyl vanadate and diethyl vanadate by the condensation of one and two molecules of ethanol, respectively, with the monoanion of V_i. It is important to note that the data indicate that for each ethanol molecule consumed in the reaction a water molecule is generated as a product. Thus the coordination number of the vanadium atom does not change. Therefore, if the V_i is tetrahedral, the ethyl esters of vanadate also are.

In the crystalline state anhydrous potassium vanadate has a four-coordinate tetrahedral structure and the monohydrate has a five-coordinate trigonal bipyramidal structure (Chasteen 1983). It has been suggested that although experimental evidence is consistent with a tetrahedral structure for the vanadate trianion in aqueous solution, the protonated forms may have a higher coordination number (Ferguson and Kustin 1979). This was based on a similar suggestion regarding molybdate, which was used to rationalize the shorter relaxation times of molybdate upon protonation of the dianion (Vold and Vold 1975). Since similar effects of pH on relaxation times of vanadate are observed, it is appropriate to question whether V_i and its esters in neutral aqueous solution actually are tetrahedral analogs of the corresponding phosphates.

Studies with V_i and the bidentate ligand ethylene glycol (Gresser and Tracey 1986) showed that the monoester and diester analogous to ethyl vanadate and diethyl vanadate formed, as well as a species giving rise to a separate resonance. Correlations of relative signal intensities with the ethylene glycol/water concentration ratio showed that the new species was formed from two V_i ions and two ethylene glycol molecules, and that three product water molecules were generated in addition to the new binuclear vanadium species. The stoichiometry of the reaction is consistent with five-coordinate vanadium atoms, if V_i is assumed to be four-coordinate. The observation of at least two signals from the new species when 1,2-propanediol or other asymmetrical bidentate ligands (Tracey and Gresser 1988) were used suggested nonequivalent environments for the two oxygens of the ligands. This was most simply rationalized in terms of a trigonal bipyramidal structure for each vanadate moiety, with each ligand bridging an apical and an equatorial coordination site on a vanadium atom, and the two vanadium atoms linked by an oxygen atom. A vanadium(V) complex with uridine is known from crystallographic studies to be five-coordinate when bound to ribonuclease A (Wlodawer et al. 1983). A non-bound uridine complex gives a [51]V nmr signal in the same chemical shift range as that of the ethylene glycol complex (Borah et al. 1985; Tracey et al. 1988).

From studies with charged bidentate ligands, such as lactate and oxalate (Ehde et al. 1986; Tracey et al. 1987), it was not possible to determine the number of product water molecules generated in forming complexes with V_i. For this reason the changes in coordination number of the complexes, relative to V_i, could not be established with the same degree of certainty as were those of the complexes formed with ethanol and ethylene glycol. However, a complex was formed which contained one bidentate lactate ligand and gave rise to a [51]V nmr signal in the same chemical shift range as the signals from the five-coordinate vicinal diol complexes. Other lactate complexes gave nmr signals in a chemical shift range separate from those in which the signals from four- and five-coordinate vanadium(V) complexes were observed. The results were consistent with six-coordinate structures for these complexes. This assignment was also consistent with the occurrence of a signal from the bis(oxalato)vanadium(V), which in crystalline form has a six-coordinate structure (Scheidt et al. 1971), in the same chemical shift range. These results, taken together, support the view that V_i and its complexes with monodentate hydroxylic ligands are tetrahedral analogs of P_i and phosphate esters, although they cannot be considered proof of this hypothesis.

Equilibrium constants for formation of monoanionic alkyl vanadate esters from V_i monoanion and alcohols ($K_f = [ROVO_3H^-]/([H_2VO_4^-][ROH])$) have values of about $0.2\ M^{-1}$, and are relatively insensitive to the pKa of the alcohol (Tracey et al. 1988), and to whether the alcohol is primary, secondary, or tertiary (Tracey and Gresser 1988). Since the pKa values of V_i and alkyl vanadates are above 8.0, this K_f value can be used to estimate the concentration of a given vanadate ester in a solution at neutral pH containing known concentrations of V_i and alcohol. This value is also reasonably close to the equilibrium constants determined for formation of other tetrahedral vanadate esters, such as those formed when the hydroxyl groups of lactate (Tracey et al. 1987) or AMP (Tracey et al. 1988) act as monodentate ligands. The equilibrium constants for formation of alkyl vanadates do not differ greatly from those for formation of alkyl phosphates, as is also true for arsenate ester formation (Long and Ray 1973; Lagunas 1980; Lagunas et al. 1984; Gresser and Tracey 1985). The equilibrium constants for formation of phenyl vanadates, as defined above for formation of alkyl vanadates, range from about 1 to 30 M^{-1}, with no simple dependence of K_f on pKa of the phenol (Galeffi and Tracey 1988). Although there is no published determination of the equilibrium constant for formation of phenyl phosphate, estimates of this value are about 10^5-fold smaller than the measured K_f for phenyl vanadate (Tracey and Gresser 1986). This could result from an interaction between the empty 3d, 4s, and 4p orbitals on vanadium and the π system of the phenol moiety in the phenyl vanadate.

Very little information is available on the rates of vanadate ester formation. Formation of vanadate esters is a likely step in the oxidation of catechols by vanadate (Ferguson and Kustin 1979), and since the oxidation is fast the ester formation must be fast. Variable temperature nmr was used to study the formation and hydrolysis of ethyl vanadate (Gresser and Tracey 1985), and it was found that the first order rate constant for hydrolysis of ethyl vanadate monoanion at neutral pH and 55° is approximately 10^8-fold larger than the corresponding value reported for ethyl phosphate at 100°. The ^{51}V nmr signal from ethyl vanadate coalesced with the V_i signal at 55°. In a study of V_i in aqueous methanol it was found that the presence of P_i in concentrations above 5mM caused coalescence of the V_i and methyl vanadate signals (Tracey et al. 1988). It has also been observed that when ligands which contained a phosphate moiety were used, separate signals were not observed for V_i and the tetrahedral vanadate esters formed from hydroxyl groups of the ligand (Tracey et al. 1988). It is therefore likely that in general, vanadate transesterification reactions come to equilibrium within milliseconds, and faster in the presence of P_i and phosphorylated compounds.

Mixed anhydrides of vanadate with P_i, As_i, pyrophosphate (Gresser et al. 1986), and with phosphate esters such as AMP (Tracey et al. 1988) have been studied by ^{51}V nmr. The formation of the mixed anhydrides is more favourable than that of phosphate anhydrides, by more than 10^6-fold in the K_f for phosphovanadate as compared to pyrophosphate at neutral pH (Gresser et al. 1986). Formation of the vanadate-vanadate anhydride divanadate is 10^8-fold more favourable than formation of pyrophosphate (Gresser et al. 1986). Formation of the ADP analog AMP-vanadate proceeds with a K_f similar to that for phosphovanadate (Tracey et al. 1988), which means that AMP-vanadate is extremely stable toward hydrolysis compared with ADP. The reasons for this very favourable formation of phosphate-vanadate mixed anhydrides are not at all clear. It may have to do with an interaction of the lone pairs of electrons on the bridging oxygen atom with orbitals on the adjacent vanadium and phosphorus atoms. If this occurred, it might be expected to cause the bridging oxygen atom to be more sp-like than sp^3, with the result that the P-O-V, or V-O-V bond angle in the anhydride would be larger than the P-O-P bond angle in phosphate-phosphate anhydrides. There does not appear to be any clear information regarding this point from crystal structures. In studies of divanadate and polyvanadates a variety of V-O-V bond angles has been reported, depending on temperature and the nature of the cation (Pope and Dale 1968; Holloway and Melnik 1986; Baglio and Dann 1972), although linear V-O-V bond angles have been observed in five-coordinate binuclear vanadium(V) complexes (Diamantis *et al.* 1986). If the V-O-V and P-O-V angles are large in divanadate and phosphovanadate, respectively, in aqueous solution, then these anhydrides might not be able to chelate Mg^{2+}, as does pyrophosphate. This appears to be the case, because although the Mg^{2+} concentration has a profoundly stabilizing effect on pyrophosphate against hydrolysis (deMeis 1984), it had no observable effect on the stability of phosphovanadate, and slightly favoured hydrolysis of divanadate at concentrations of $MgCl_2$ up to 30mM (Craig and Gresser 1989).

The rates of formation and hydrolysis of phosphovanadate anhydrides have not been determined. They are evidently faster than those of vanadate esters, because the anhydrides are in fast chemical exchange with their hydrolysis products. It was possible to determine the stoichiometries and equilibrium constants for some of the anhydride formation reactions in two ways. The change in chemical shift of the average V_i plus anhydride signal as the phosphate concentration varied was analyzed quantitiatively (Gresser *et al.* 1986). Another method which has been employed in a number of systems was to make use of the facts that the signal from the vanadate tetramer can often be integrated accurately, and the tetramer appears not to interact with many of the ligands studied. Thus, it is possible to determine the concentration of the tetramer, and therefore the concentration of V_i, since the equilibrium constant for interconversion of V_i and the tetramer can be determined under the appropriate conditions. Thus, even when a complex gives a signal which cannot be resolved from the V_i signal, the vanadium atom concentrations of V_i and the other species can each be determined by making use of the V_i/V_4 equilibrium (Gresser *et al.* 1986; Tracey *et al.* 1987).

In studies of V_i in aqueous methanol (Tracey *et al.* 1988) it was found that methyl esters of divanadate can form spontaneously, with equilibrium constants similar to those for formation of esters of monovanadate. Evidence for condensation of divanadate with other hydroxyl groups has also been obtained (Tracey and Gresser, 1988). Similarly, the ATP analog AMP-divanadate can form by condensation of divanadate with the phosphate of AMP (Tracey *et al.* 1988). The equilibrium constant for formation of this mixed anhydride was found to be similar to that for formation of mixed phosphate-vanadate anhydrides with V_i and P_i or pyrophosphate.

It is clear, from ^{51}V nmr studies, that vanadate anhydrides analogous to nucleoside phosphates form spontaneously in aqueous solution. The chemical shift range in which the vanadium nmr signals of these species occur suggests that the vanadate moieties are tetrahedral, or at least have structures similar to V_i and divanadate. However, the extraordinary stability, relative to phosphate-phosphate anhydrides, of phosphovanadate anhydrides toward hydrolysis, along with the apparent inability of phosphovanadate to chelate Mg^{2+}, suggests that significant structural differences exist between these two types of anhydrides. These differences may well be the reason why phosphovanadate anhydrides have been found to interact strongly with only a very few enzymes. It is also possible that investigators have simply not yet looked hard enough for such interactions. This point is discussed further below.

Vanadates as Inhibitors of Enzymatic Phosphoryl Transfer

A large body of evidence indicates that enzymatic phosphoryl transfer reactions proceed via transition states in which the phosphorus atom of the transferred phosphoryl group is five-coordinate (Knowles 1980). This is a very high energy structure for phosphate, and phosphoryl transfer enzymes catalyse their reactions by stabilizing this structure by strong binding interactions. Since vanadate can adopt five-coordinate structures readily, as shown by studies of crystalline (Chasteen 1983) and aqueous solution (Gresser and Tracey 1986) vanadium(V) complexes, it was reasonable to expect that some vanadate complexes would bind very tightly to some phosphoryl transfer enzymes, thus inhibiting them even at very low concentrations. This was the reasoning of Lienhard and coworkers when they initiated a study of the inhibition of ribonuclease A by uridine-vanadate (Lindquist et al. 1973). It was found that V_i itself was a poor inhibitor of RNase, but the uridine-vanadate complex was very potent. The inhibitory complex was assumed to have a structure in which a five-coordinate vanadium(V) atom has two of its coordination sites occupied by the 2 and 3 hydroxyls of the uridine ligand. This has been confirmed by X-ray and neutron diffraction studies of RNase with the bound inhibitor (Wlodawer et al. 1983). A dissociation constant of 1.2×10^{-5} M was determined for uridine-vanadate bound to RNase A (Lindquist et al. 1973). This is almost certainly an overestimate of the dissociation constant, because the major uridine-vanadate complex which forms in solution is not the one which binds to RNase (Tracey et al. 1988). The main complex formed is a binuclear vanadium(V) complex in which the vanadium atoms are each thought to be five-coordinate and have a bidentate uridine ligand. It has not yet been possible to detect in solution a mononuclear five-coordinate uridine-vanadate complex, or to estimate an equilibrium constant for its formation. Therefore it appears that uridine-vanadate is a better transition-state analog for the RNase catalysed hydrolysis of uridine $2',3'$-phosphate than it has been thought to be.

An early study by Van Etten and coworkers focused on the inhibition of acid phosphatase by vanadate and other transition metal oxyanions (Van Etten et al. 1974). Here the reasoning was that enzyme-bound V_i could easily take on a structure resembling the transition-state for transfer of the phosphate of the phosphorylated enzyme intermediate to water in the normal catalytic reaction. Strong inhibition of acid phosphatase was observed, with K_i values below 10^{-6} M. Again in this instance, the kinetically determined K_i probably underestimates how good a transition-state analog the enzyme-bound vanadate really is, because the bound vanadate very likely has a different structure from the tetrahedral V_i in solution, still assuming that aqueous V_i is tetrahedral.

In spite of these interesting and valuable studies, biochemists were not particularly interested in vanadate until the discovery by Cantley and coworkers that it was the contaminant in commercial ATP which strongly inhibited the Na^+K-ATPase (Cantley et al. 1977). The discovery of the insulin mimetic effects of vanadium further increased interest (Dubyak and Kleinzeller 1980; Shechter and Karlish 1980). Since these reports the effects of vanadate on a number of phosphoryl transfer enzymes have been studied, and summaries of some of these can be found in recent reviews (Chasteen 1983; Nechay et al. 1986). In general, enzymes whose catalytic mechanisms involve hydrolysis of a phosphorylated enzyme intermediate, such as the ion translocating ATPases (Cantley et al. 1977), acid phosphatases (Van Etten et al. 1974; Stankiewicz and Gresser 1988), alkaline phosphatase, and aryl sulfatase (Stankiewicz and Gresser 1988) are strongly inhibited by V_i. Enzymes which do not catalyse the transfer of phosphate from themselves to water, and do not use nucleoside phosphate substrates, such as phosphoglucomutase (Percival et al. 1989 sub.) phosphoglycerate mutase (Liu 1989), and ribonuclease (Lindquist et al. 1973), are not significantly inhibited by V_i alone, but are very strongly inhibited by vanadate complexes which can be transition-state analogs for reaction steps of the normal catalytic mechanism. This is discussed further below. Enzymes for which nucleoside disphosphates or triphosphates are substrates, and which do not transfer phosphate from phospho-enzyme intermediates to water, such as kinases (Craig and Gresser 1989; Climent et al. 1981) and the mitochondrial proton-translocating ATPase (Bramhall 1987; O'Neal et al. 1979) are not strongly inhibited by either V_i or its complexes. Known exceptions to this last generalization include the myosin ATPase (Goodno 1979) and the dyenin ATPase (Gibbons et al. 1978). Very likely more exceptions to this and the other two generalizations made above will emerge as the number of detailed studies of inhibition of phosphoryl transfer enzymes by V_i increases. At present, however, these general statements appear to provide a useful way of looking at the field. These three general categories of vanadate inhibition will be discussed briefly.

The case of enzymes which, in one step of their catalytic mechanism, transfer a covalently bound phosphate group to water is straightforward. The only components involved in the transition state are the phosphate group, the solvent, and functional groups of the enzyme. It is therefore to be expected that V_i can strongly inhibit enzymes of this class by itself, by spontaneously becoming covalently linked to the functional group which is phosphorylated in the normal catalytic reaction, and then forming a stable, five-coordinate structure resembling the transition-state for the dephosphorylation step of the phosphatase reaction. However, formation of the phosphorylated enzyme intermediate is another step of the catalytic mechanism of these enzymes. Binding interactions with parts of the phosphoryl donor other than the phosphate group can be important both in determining the specificity of the enzyme and in increasing the reaction rate by transition-state stabilization. When such interactions are important in the phosphorylation step of the phosphatase reaction, a complex of vanadate with the dephosphorylated substrate can be a stronger inhibitor than vanadate by itself. This case is analogous to the inhibition of RNase by the uridine-vanadate complex (Lindquist et al. 1973). This effect can be important even with enzymes for which the rate-limiting step at saturating substrate concentrations is hydrolysis of the phosphoenzyme intermediate, because the phosphorylation step is also a reaction which is catalysed by tight binding of its transition-state.

In terms of inhibition of phosphoprotein phosphatases under physiological conditions, where the dephosphorylated substrate of the phosphatase reaction can be present at significant concentrations, this effect is of considerable potential importance, and has been discussed in detail in a recent review (Gresser *et al.* 1987). It provides a possible rationalization for the selective inhibition of protein phosphatases by V_i. For cases in which strong binding interactions between the catalytic site of the protein phosphatase enzyme and the protein portion of its phosphoprotein substrate are important for its specificity and catalysis, strong inhibition by V_i might actually be due mainly to a vanadate-protein complex. Inhibition by V_i itself might not be strong enough to be significant under physiological conditions. In the case of protein phosphatases whose specificity is determined largely by other factors, such as the "targeting subunits" discussed recently by P. Cohen (Cohen 1988; Cohen 1989), presence of the dephosphorylated phosphoprotein substrate might not enhance inhibition by V_i, and V_i might not be a significant inhibitor of these phosphatases under physiological conditions. Certainly some phosphatases appear to be quite nonspecific, catalysing hydrolysis of a wide range of phosphorylated substrates with similar k_{cat}/Km values, while others are very sensitive to substrate structure (Hall and Williams 1986; Stankiewicz and Gresser 1988). In model studies with commercially available phosphatases, one case was found in which inhibition by V_i was enhanced by added dephosphorylated substrate, and another case was found in which added dephosphorylated substrate had no effect on inhibition by V_i (Stankiewicz and Gresser 1988).

In the second general category of phosphoryl transfer enzymes inhibited by vanadate, those which do not catalyse the hydrolysis of a phosphoenzyme intermediate, and do not take nucleoside phosphates as substrates, it would be surprising if V_i by itself, were a potent inhibitor. In the case of an enzyme like RNase, which does not form a phosphoenzyme intermediate, V_i alone could not mimic an intermediate or transition-state by binding directly to the enzyme. The other parts of the transition-state structure, those which are covalently attached to the phosphate moiety, are needed to complete the transition-state analog structure. This interaction of vanadate, with uridine in the case of the uridine-vanadate RNase inhibitor, is needed to form links between the vanadate moiety and the enzyme, and to hold the vanadate moiety in the correct position relative to those enzyme functional groups which interact with it directly. This function of the uridine moiety is obvious.

Another role of the ligand of the inhibitory vanadate complex may well be to furnish binding energy which is used to change the enzyme to a state in which it interacts more strongly with the vanadate transition-state analog moiety. One of the clearest examples of this type of behaviour in other systems was provided by studies of elastase, in which an aldehyde moiety was placed at the position of the scissile amide bond in peptide substrate analogs (Thompson 1973; Thompson 1974). The hydrated aldehyde moiety acts as a transition-state analog, and its binding constant, relative to that of the corresponding peptide substrate, is increased by a factor of nearly 100 by the addition of an amino acid four residues removed from the position of the scissile bond. It is likely that this effect of a change in the substrate at a position remote from the site of the chemical reaction is communicated through the protein, rather than by simply furnishing a better link between the transition-state analog moiety and the enzyme (Jencks 1975).

In the case of nonhydrolytic phosphoryl transfer enzymes which do form phosphoryl enzyme intermediates, hydrolysis of the phosphoryl enzyme is an undesirable side reaction. Therefore it would be very surprising if an analog of the transition-state for this reaction bound tightly to the enzyme. Examples of such enzymes are the phosphomutases, phosphoglycerate mutase and phosphogluco-mutase. Although these enzymes are both known to be inhibited by vanadate (Climent et al. 1981; Ninfali et al. 1983; Carreras et al. 1988; Carreras et al. 1982), it was not appreciated until recently that the actual inhibitor of these enzymes is not V_i, but a complex of V_i with the substrate (Percival et al. 1989 sub.; Liu et al. 1988). It is understandable that this was not apparent in initial studies, because of the way in which the experiments were done, and because of the catalytic mechanism of these enzymes.

For both enzymes the reaction proceeds by binding of one monophosphate isomer (3-phosphoglycerate or glucose 1-phosphate) to the phosphoenzyme, followed by phosphoryl transfer to yield the diphosphorylated intermediate (2,3-diphosphoglycerate or glucose 1,6-diphosphate) bound to the dephospho form of the enzyme, followed by phosphoryl transfer to yield the isomerized mono-phosphate bound to the phosphoenzyme. The product (2-phosphoglycerate or glucose 6-phosphate) then dissociates to regenerate the free phosphoenzyme. The bound diphosphorylated intermediate can dissociate to yield the inactive de-phosphoenzyme, so the diphosphorylated species must be present in the reaction mixture to keep the enzyme active. This mechanism is of the ping-pong type, in which the diphosphorylated intermediate is both the first product and the second substrate. The mixed phosphate, vanadate diesters glucose 1-phosphate 6-vanadate, and 2-vanado 3-phosphoglycerate can bind to the dephospho form of the appropriate enzyme, and act as transition-state analogs for the first of the phosphoryl transfer steps mentioned above. Kinetic studies of phosphogluco-mutase show that glucose 1-phosphate 6-vanadate binds to the dephospho form of phosphoglucomutase with a dissociation constant of $10^{-12}M$, while there is no detectable inhibition ascribable to V_i alone (Percival et al. 1989 sub.). Kinetic and equilibrium binding studies of phosphoglycerate mutase yielded qualitatively similar results (Liu et al. 1988; Liu 1989). In previous kinetic studies the substrate whose concentration was varied was the diphosphate cofactor. The concentration of the monophosphate substrate was not varied, so it was not recognized that the inhibition was totally dependent upon the presence of the monophosphate sub-strate, and became stronger at higher concentrations of this substrate.

It is reasonable to anticipate that other enzymes of this general type will also be insensitive to inhibition by V_i alone, but a larger number of detailed studies should be done. At present this generalization, like the others presented here, simply provides a working hypothesis.

The third generalization has to do with the very weak inhibition by V_i of kinases (Craig and Gresser 1989, Climent et al. 1981), and ATPases which do not form phosphoenzyme intermediates (Bramhall 1987; O'Neal et al. 1979). In this case the generalization is already flawed, and thus made more interesting, by the exceptions which are strongly inhibited by V_i. These are the myosin ATPase (Goodno 1979) and the dynein ATPase (Gibbons et al. 1978). The actual inhibitor in these cases appears to be an ADP-vanadate complex, which functions as a transition-state analog for the ATP hydrolysis reaction. Advantage has been taken of the extremely tight binding of this complex to achieve very specific labelling of myosin by photoaffinity analogs of ADP (Munson et al. 1986), and to chemically modify myosin at the catalytic site by vanadate-sensitized photochemical reactions, including photocleavage of specific peptide bonds (Grammer et al. 1988; Cremo et al. 1988). The vanadium-sensitized photocleavage reaction had previously been used by I.R. Gibbons in studies of the dynein ATPase (Lee-Eiford et al. 1986; Gibbons et al. 1987; Gibbons and Gibbons 1987; Tang and Gibbons 1987). In the case of myosin, a myosin-Mg^{2+}-ADP-vanadate complex forms, and is so stable that nonbound Mg^{2+}-ADP-vanadate can be removed by gel filtration. The bound vanadate complex dissociates from the myosin over a period of days (Grammer et al. 1988).

It was to be expected, for reasons discussed earlier in this chapter, that ADP-vanadate would act as a tight-binding transition-state analog inhibitor toward myosin and other enzymes which catalyse transfer of the gamma phosphate of ATP to water or other acceptors. It is therefore surprising that V_i is such a poor inhibitor of the kinases. It was considered that the reason for the weak inhibition of kinases by V_i may have had to do with the conditions used in the experiments. For example, hexokinase would normally be studied in the presence of the substrates ATP and glucose, and V_i would be added as an inhibitor. The appropriate transition-state analog would be a complex in which ADP and glucose are both apical ligands to a five coordinate trigonal bipyramidal vanadium(V) atom. Thus, inhibition by vanadate should be stronger in the presence of added ADP. This prediction was tested for several kinases, including hexokinase and adenylate kinase (Craig et al. 1988), with negative results. One possible explanation for these results arises from the ^{51}V nmr studies discussed earlier, which indicate that the phospho-vanadate anhydride does not chelate Mg^{2+}, as does pyrophosphate. Since a number of enzymes for which Mg^{2+}-ATP is a substrate are known to bind the complex in which the Mg^{2+} is chelated by the beta and gamma phosphates of ATP (Mildvan and Fry 1987), ADP-vanadate might not bind to these enzymes because it cannot chelate Mg^{2+} correctly. Another possibility is that, assuming the P-O-V bond angle of the phosphovanadate anhydride moiety of the ADP-vanadate is nearly linear, the complex simply will not fit into the catalytic site correctly. In the transition-state analog, the O-V-O bond angle of the system involving the apical oxygen ligands from the ADP and the phosphate acceptor should be 180°. If the P-O-V bond angle of the phosphovanadate anhydride moiety is also 180°, the complex might be too different from the actual transition-state to be a good transition-state analog.

The reason for the tight binding of ADP-vanadate to the myosin and dynein ATPases could be that, 1) the Mg^{2+} is not chelated by the beta and gamma phosphates of ATP in the normal reaction, 2) the binding of the complex is strong enough to change the P-O-V bond angle so that Mg^{2+} can be chelated correctly, 3) the longer structure resulting from the larger P-O-V bond angle can be accommodated by the catalytic sites of these ATPases, or 4) some other reason. Reason number four is probably the most likely explanation. It is too early in studies of the interactions of vanadate with this class of ATP utilizing enzymes to try to discern a pattern, but perhaps it will be useful at this stage to think of these enzymes as a class, distinct from the other two classes discussed earlier.

Other Interactions of Vanadate with Phosphoryl Transfer Enzymes

There are a number of reports of effects attributed to interaction of vanadate oligomers with enzymes. These include inhibition of phosphofructokinase by decavanadate (Choate and Mansour 1979), binding of decavanadate and other oligomers to the sarcoplasmic reticulumn ATPase (Csermely et al. 1985; Coan et al. 1986), inhibition by decavanadate of hexokinase, adenylate kinase, and phosphofructokinase (Boyd et al. 1985), activation of the 2,3-diphosphoglycerate phosphatase activity of phosphoglycerate mutase by divanadate (Stankiewicz et al. 1987), photocleavage of dynein ATPase sensitized by oligovanadate (Tang and Gibbons 1987), inhibition of transducin GTPase by decavanadate (Kanaho et al. 1988), and binding of the tetramer to myosin S-1 (Cremo et al. 1988). In most of these cases, it is not clear whether the vanadate oligmer is binding as an analog of a phosphate, or simply as a rather large polyanion binding at sites whose normal ligands are large polyanions. In the case of the divanadate-activated 2,3-diphosphoglycerate phosphatase activity of phosphoglycerate mutase, the evidence supports binding of divanadate at the catalytic site of the phosphoenzyme intermediate as an analog of the substrate. The result is apparently that the phosphoenzyme is "tricked" into transferring its phosphate to water, and is rephosphorylated by 2,3-diphosphoglycerate. It has recently been found that the vanadate tetramer binds to the dephospho form of phosphoglucomutase with a dissociation constant of $5x10^{-7}M$ (Percival and Gresser unpublished results).

There are a few reports of activating effects of vanadate. These include activation of adenylate cyclase (Schwabe et al. 1979), and activation of the tyrosine kinase activities of various peptide hormone receptors and oncogene products (Gresser et al. 1987). The mechanism of activation of the adenylate cyclase is not clear, but it could be due to interference with some regulatory process involving phosphoryl transfer, such as protein phosphorylation or the GTPase activity of a G protein. Activation of the receptor tyrosine kinase by vanadate very likely is due to interference with the phosphorylation of specific tyrosine residues on the receptor. It can safely be assumed that vanadate will spontaneously esterify the tyrosines which normally become phosphorylated upon binding of the hormone. It is quite reasonable to further assume that esterification by vanadate will activate the tyrosine kinase activity of the receptor, as does esterification by phosphate (Tracey and Gresser 1986). It is not known whether this activation mechanism is sufficient to account for the physiological effects of vanadate, such as its insulin-mimetic effect, on tyrosine kinases. It is considered more likely that the tyrosine kinase activation results from inhibition of receptor dephosphorylation by phosphotyrosine protein phosphatases (Gresser *et al.* 1987). Some of the other activating effects of vanadate, such as the 200-fold enhancement of the phosphorylation of glucose by the phospho form of phosphoglucomutase (Layne and Najjar 1979), can be explained by spontaneous formation of vanadate esters which act as substrates, or by V_i acting as a nucleophile toward an enzyme-bound intermediate. Other activating effects, such as that on the 2,3-diphosphoglycerate phosphatase activity of phosphoglycerate mutase (Stankiewicz *et al.* 1987), will be found to be due to specialized mechanisms, rather than to a general type of behaviour of vanadate.

Conclusions

Vanadate can spontaneously form esters which resemble the analogous phosphate esters. This conclusion is supported by studies using enzyme kinetics and ^{51}V nmr structural, equilibrium, and kinetic studies. The vanadate esters can act as substrates for enzymes which do not catalyse phosphoryl transfer reactions and as inhibitors of some enzymes which do catalyse phosphoryl transfer reactions. We suggest that it is useful to consider three general classes of phosphoryl transfer enzymes when thinking about their interactions with vanadates: 1) Enzymes which catalyse the transfer of phosphate from a phosphoenzyme intermediate to water (inhibited by V_i alone, but inhibition by vanadate complexes can be stronger) 2) Enzymes which do not catalyse the hydrolysis of a phosphoenzyme intermediate, other than those which catalyse reactions of nucleoside phosphates (not inhibited by V_i alone, but inhibited by vanadate complexes), and 3) Enzymes which catalyse reactions of nucleoside phosphates (generally not strongly inhibited by V_i or vanadate complexes, with some exceptions.)

Finally, it must be realized that only vanadium (V) has been considered in this chapter. The physiological effects of vanadium (IV) and vanadium (III) may be of equal or greater significance than those of vanadium (V). The aqueous chemistry and the biochemistry of vanadium is still very little explored, and poorly understood. It is expected that many interesting and significant surprises will be reported as the field develops. We hope that some of the investigators who make these future discoveries will have found this review useful.

References

Adams, S.R., Sparkes, M.J. and Dixon, H.B.F. (1984) Biochem. J. **221**, 829–836.
Baes, C.F., Jr. and Mesmer, R.E. (1976) The Hydrolysis of Cations, Wiley, London.
Baglio, J.A. and Dann, J.N. (1972) J. Solid State Chem. **4**, 87–93.
Borah, B., Chen, C., Egan, W., Miller, M., Wlodawer, A. and Cohen, J.S. (1985) Biochemistry **24**, 2058–2067.

Boyd, D.W., Kustin, K. and Niwa, M. (1985) Biochim. Biophys. Acta. **827**, 472–475.

Bramhall, E.A. (1987) M.Sc. Thesis, Simon Fraser University.

Cantley, L.C., Jr., Josephson, L., Warner, R., Yanagisawa, M., Lechene, C. and Guidotti, G. (1977) J. Biol. Chem. **252**, 7421–7423.

Carreras, J., Climent, F., Bartrons, R. and Pons, G. (1982) Biochim. Biophys. Acta **705**, 238–242.

Carreras, J., Bartrons, R., Grisolia, S. (1980) Biochem. Biophys. Res. Commun. **96**, 1267–1273.

Carreras, M., Bassols, A.M., Carreras, J. and Climent, F. (1988) Arch. Biochem. Biophys. **264**, 155–159.

Chasteen, N.D. (1981) Biol. Magn. Reson. **3**, 53–119.

Chasteen, N.D. (1983) Struct. Bonding **53**, 105–138.

Choate, G. and Mansour, T.E. (1979) J. Biol. Chem. **254**, 11457–11462.

Climent, F., Bartrons, R., Pons, G. and Carreras, J. (1981) Biochem. Biophys. Res. Commun. **101**, 570–576.

Coan, C., Scales, D.J. and Murphy, A.J. (1986) J. Biol. Chem. **261**, 10394–10403.

Cohen, P. (1989) Annu. Rev. Biochem. **58**, 453–508.

Cohen, P. (1988) J. Cell. Biol. **107**, 227a.

Craig, M.M. and Gresser, M.J. (1989) J. Cell. Biol. **107**, 189a.

Craig, M.M. and Gresser, M.J. (unpublished results).

Craig, M.M. (1986) M. Sc. Thesis, Simon Fraser University.

Cremo, C., Grammer, J., Long, G.T. and Yount, R. (1988) J. Cell. Biol. **107**, 257a.

Cremo, Cr.R., Grammer, J.C. and Yount, R.G. (1988) Biochemistry **27**, 8415–8420.

Csermely, P., Martonosi, A., Levy, G.C. and Ejchart, A.J. (1985) Biochem. J. **230**, 807–815.

de Meis, L. (1984) J. Biol. Chem. **259**, 6090–6097.

DeMaster, E.G. and Mitchell, R.A. (1973) Biochemistry **12**, 3616–3621.

Diamantis, A.A., Frederiksen, J.M., Salam, M.A., Snow, M.R. and Tiekink, E.R.T. (1986) Aust. J. Chem. **39**, 1081–1088.

Doudoroff, M., Barker, H.A. and Hassid, W.Z. (1947) J. Biol. Chem. **170**, 147–150.

Drueckhammer, D.G., Durrwachter, J.R., Pederson, R.L., Crans, D.C., Daniels, L. and Wong, C.-H (1989) J. Org. Chem. **54**, 70–77.

Dubyak, G.R. and Kleinzeller, A. (1980) J. Biol. Chem. **255**, 5306–5312.

Ehde, P.M. andersson, I. and Petterson, L. (1986) Acta Chem. Scand. **A40**, 489–499.

Erdmann, E., Werdan, K., Krawietz, W., Schmitz, W. and Scholz, H. (1984) Biochem. Pharmacol. **33**, 945–950.

Ferguson, J.H. and Kustin, K. (1979) Inorg. Chem. **18**, 3349–3357.

Galeffi, B. and Tracey, A.S. (1988) Can. J. Chem. **66**, 2565–2569.

Gibbons, B.H. and Gibbons, I.R. (1987) J. Biol. Chem. **262**, 8354–8359.

Gibbons, I.R., Cosson, M.P., Evans, J.A., Gibbons, B.H., Houck, B., Martinson, K.H., Sale, W.S. and Tang, W.J.Y. (1978) Proc. Natl. Acad. Sci. U.S.A. **75**, 2220–2224.

Gibbons, I.R., Lee-Eiford, A., Mocz, G., Phillipson, C.A., Tang, W.-J.Y. and Gibbons, B.H. (1987) J. Biol. Chem. **262**, 2780–2786.

Goodno, C.C. (1979) Proc. Natl. Acad. Sci. U.S.A. **76**, 2620–2624.

Grammer, J.C., Cremo, C.R. and Yount, R.G. (1988) Biochemistry **27**, 8408–8415.

Gresser, M.J. and Tracey, A. (1986) J. Am. Chem. Soc. **108**, 1935–1939.

Gresser, M.J. Tracey, A.S. and Stankiewicz, P.J. (1987) Adv. Protein Phosphatases **4**, 35–57.

Gresser, M.J., Tracey, A.S. and Parkinson, K.M. (1986) J. Am. Chem. Soc. **108**, 6229–6234.

Gresser, M.J. (1981) J. Biol. Chem. **256**, 5981–5983.

Gresser, M.J. and Tracey, A. (1985) J. Am. Chem. Soc. **107**, 4215–4220.

Guranowski, A. and Blanquet, S. (1986) Biochimie **68**, 757–760.

Hall, A.D. and Williams, A. (1986) Biochemistry **25**, 4784–4790.

Hathcock, J.N., Hill, C.H. and Tove, S.B. (1966) Can. J. Biochem. **44**, 983–988.

Holloway, C.E. and Melnik, M. (1986) Rev. in Inorg. Chem. **8**, 287–360.

Itada, N. and Cohn, M. (1963) J. Biol. Chem. **238**, 4026–4031.

Jacobson, S.E. Mares, F. and Zambri, P.M. (1979) a: J. Am. Chem. Soc. **101**, 6938–6946.

Jacobson, S.E. Mares, F. and Zambri, P.M. (1979) b: J. Am. Chem. Soc. **101**, 6946–6950.

Jandhyala, B.S. and Hom, G.J. (1983) Life Sci. **33**, 1325–1340.

Jencks, W.P. (1975) Adv. Enzymol. Relat. Areas Mol. Biol. **43**, 219–410.
Kanaho, Y., Chang, P.P., Moss, J. and Vaughn, M. (1988) J. Biol. Chem. **263**, 17584–17589.
Kepart, D.L. (1973) Compr. Inorg. Chem. **4**, 607–672.
Knowles, J.R. (1980) Ann. Rev. Biochem. **49**, 877–919.
Kustin, K, McLeod, G.C., Gilbert, T.R. and Briggs, L.B.R., 4th (1983) Struct. Bonding **53**, 139–160.
Lagunas, R. (1980) Arch. Biochem. Biophys. **205**, 67–75.
Lagunas, R., Pestaña, D. and Diez-Masa, J.C. (1984) Biochemistry **23**, 955–960.
Layne, P.P. and Najjar V.A. (1979). Proc. Acad. Sci. U.S.A. **76**, 5010–5013.
Lee-Eiford, A., Ow, R.A. and Gibbons, I.R. (1986) J. Biol. Chem. **261**, 2337–2342.
Lindquist, R.N., Lynn, J.L., Jr. and Lienhard, G.E. (1973) J. Am. Chem. Soc. **95**, 8762–8768.
Liu, S. (1989) M. Sc. Thesis, Simon Fraser University.
Liu, S., Stankiewicz, P.J., Gelb, M., Black, E., Tracey, A.S. and Gresser, M.J. (1988) J. Cell Biol. **107**, 189 a.
Long, J.W. and Ray, W.J., Jr. (1973) Biochemistry **12**, 3932–3937.
Mildvan, A.S. and Fry, D.C. (1987) Adv. Enzymol. **59**, 241–313.
Moore, S.A., Moennich, D.M.C. and Gresser, M.J. (1983) J. Biol. Chem. **258**, 6266–6271.
Munson, K.B., Smerdon, M.J. and Yount, R.G. (1986) Biochemistry **25**, 7640–7650.
Nechay, B.R. (1984) Annu. Rev. Pharmacol. Toxicol. **24**, 501–524.
Nechay, B.R., Nanninga, L.B., Nechay, P.S.E., Post, R.L., Grantham, J.J., Macara, I.G., Kubena, L.F., Phillips, T.D. and Nielsen, F.H. (1986) Fed. Proc. **45**, 123–132.
Ninfali, P., Accorsi, A., Fazi, A., Palma, F. and Fornaini, G. (1983) Arch. Biochem. Biophys. **226**, 441–447.
Nour-Eldeen, A.F., Craig, M.M. and Gresser, M.J. (1985) J. Biol. Chem. **260**, 6836–6842.
O'Neal, S.G., Rhoads, D.B. and Racker, E. (1979) Biochem. Biophys. Res. Commun. **89**, 845–850.
Percival, M.D., Doherty, K. and Gresser, M.J. (1989) submitted.
Percival, M.D. and Gresser, M.J. unpublished results.
Pope, M.T. and Dale, B.W. (1968) Q. Rev. Chem. Soc. **22**, 527–548.
Ramasarma, T and Crane, F.L. (1981) Curr. Top. Cell. Regul. **20**, 247–301.
Rehder, D. (1977) Z. Naturforsch., B; Anorg. Chem., Org. Chem. **326**, 771–775.
Rehder (1982) Bull. Magn. Reson. **4**, 33–83.
Rozovskaya, T.A., Rechinsky, V.O., Bibilashvili, R.S., Karpeisky, M.Y., Tarusova, N.B., Khomutov, R.M. and Dixon, H.B.F. (1984) Biochem. J. **224**, 645–650.
Scheidt, W.R., Tsai, C.-C. and Hoard, J.L. (1971) J. Am. Chem. Soc. **93**, 3867–3872.
Schwabe, U., Puchstein, C., Hannemann, H. and Söchtig, E. (1979) Nature **277**, 143–145.
Shechter, Y. and Karlish, S.J.D. (1980) Nature **284**, 556–558.
Slocum, D.H. and Varner, J.E. (1960) J. Biol. Chem. **235**, 492–495.
Slooten, L. and Nuyten, A. (1983) Biochim. Biophys. Acta **725**, 49–59.
Slooten, L. and Nuyten, A. (1984) Biochim. Biophys. Acta **766**, 88–97.
Stankiewicz, P.J. Gresser, M.J., Tracey, A.S. and Hass, L.F. (1987) Biochemistry **26**, 1264–1269.
Stankiewicz, P.J. and Gresser, M.J. (1988) Biochemistry **27**, 206–212.
Tang, W-J.Y. and Gibbons, I.R. (1987) J. Biol. Chem. **262**, 17, 17728–17734.
Thompson, R.C. (1973) Biochemistry **12**, 47–51.
Thompson, R.C. (1974) Biochemistry **13**, 5495–5501.
Tracey, A.S., Gresser, M.J. and Liu, S. (1988). J. Am. Chem. Soc. **110**, 5869–5874.
Tracey, A.S., Gresser, M.J. and Parkinson, K.M. (1987) Inorg. Chem. **26**, 629–638.
Tracey, A.S. and Gresser, M.J. (1988) Can. J. Chem. **66**, 2570–2574.
Tracey, A.S., Gresser, M.J. and Galeffi, B. (1988) Inorg. Chem. **27**, 157–161.
Tracey, A.S. and Gresser, M.J. (1988) Inorg. Chem. **27**, 2695–2702.
Tracey, A.S. and Gresser, M.J. (1986) Proc. Natl. Acad. Sci. U.S.A. **83**, 609–613.
Tracey, A.S., Galeffi, B. and Mahjour, S. (1988) Can. J. Chem. **66**, 2294–2298.
VanEtten, R.L., Waymack, P.P. and Rehkop, D.M. (1974) J. Am. Chem. Soc. **96**, 6782–6785.
Visedo-Gonzalez, E. and Dixon, H.B.F. (1989) Biochem. J. **260**, 299–301.
Vold, R.R. and Vold, R.L. (1975) J. Magn. Res. **19**, 365–371.
Warburg, O. and Christian, W. (1939) Biochem. Z. **303**, 40–68.
Wlodawer, A., Miller, M. and Sjölin, L. (1983) Proc. Natl. Acad. Sci. U.S.A. **80**, 3628–3631.

V. Vanadium Haloperoxidases

R. WEVER and B.E. KRENN

E.C. Slater Institute for Biochemical Research and Biotechnological Center, University of Amsterdam, Plantage Muidergracht 12, 1018 TV Amsterdam, The Netherlands.

Introduction

Since the beginning of the 19th century peroxidases have attracted the attention of chemists, biologists and physiologists. These oxido-reductases catalyze the oxidation of a variety of organic as well as inorganic electron donors such as Cl^-, Br^-, I^- or SCN^- by hydrogen peroxide (H_2O_2) or other hydroperoxides (ROOH). The overall catalyzed reaction for an organic electron donor is:

$$H_2O_2 + 2AH_2 \rightarrow 2AH^\circ + 2H_2O \qquad (1)$$

where AH_2 is the electron donor and AH° the oxidized product which may react further. In this reaction, one electron step oxidations (Chance, 1949) occur. In general the products are highly colored and the reaction can be monitored by eye.

Thus, the first recorded peroxidase reaction dated back as early as 1809 when guaiacum was used as a dentifrice, and converted into an intense blue substance (see Saunders *et al.*, 1964). However, the presence of a peroxidase in human saliva was not reported until 1909 (MacDonald *et al.*, 1909). In 1810, Planche noticed the same color reaction when guaiacum was added to fresh roots of horseradish and in 1855, Schönbein recorded that the oxidation of certain organic compounds by hydrogen peroxide could be catalyzed by extracts of plant as well as animal tissues. Linossier (1898) was the first who demonstrated that the oxidation of organic compounds (observed in pus) was catalyzed by a peroxidase, an enzyme quite distinct from the already known oxidases. In the first decade of this century, Bach and Chodat (1903) emphasized the wide occurrence of peroxidases in plants, whereas Batelli and Stern (1908) demonstrated the presence of these enzymes in many animal tissues. Various algae, which are low in the scale of plant development have also been shown to contain peroxidases (Atkins, 1914; Reed, 1915; Tamiya, 1934; Petersson, 1940; Rönnerstrand, 1946).

As indicated peroxidases are also able to oxidize inorganic electron donors such as halides. The reaction mechanism differs from that mentioned above, a two-electron oxidation of the donor (X^-) occurs rather than two one-electron consecutive steps (Eqn. 2).

$$H_2O_2 + X^- + H^+ \rightarrow HOX + H_2O \qquad (2)$$
$$HOX + AH \rightarrow H_2O + AX \qquad (3)$$

The reaction products are hypohalous acids (HOX) and if a nucleophilic acceptor (AH) is present halogenated compounds (AX) are produced (Eqn. 3). Most

N. Dennis Chasteen (ed.), Vanadium in Biological Systems, 81–97.

peroxidases are able to carry out both reactions, and enzymes, able to carry out the latter reaction, are collectively called haloperoxidases. This distinction, however, is sometimes artificial. Most peroxidases do oxidize halides, but the distinct ability to oxidize chloride is limited to a few peroxidases such as chloroperoxidase (Hager et al., 1966) and myeloperoxidase (Harrison and Schultz, 1976; Bakkenist et al., 1980). In contrast, some peroxidases are very specific, they are only able to oxidize bromide or iodide and are unable to directly use organic electron donors such as guaiacol and O-dianisidine (De Boer et al., 1987; Wever et al., 1985b). However, these compounds will of course be oxidized by the hypohalous acid being produced.

Until the eighties most of the peroxidases were known to contain a heme prosthetic group and for the properties of these enzymes the interested reader is referred to the reviews by Hewson and Hager (1979), Paul (1963), Dunford and Stillman, (1976), Morrison and Schonbaum (1976) or more recently Neidleman and Geigert (1986). Lately, it has been shown that the presence of a heme group is no absolute requirement for enzymatically active haloperoxidases. This was not totally unexpected, since already some non-halogenating peroxidases without a heme group were known, e.g. glutathione peroxidase which contains selenium as prosthetic group (Ladenstein and Wensel, 1976) and a flavoperoxidase from Streptoccocus faecalis (Dolin, 1957). Murphy and O'hEocha (1973) purified an iodoperoxidase from the green alga Enteromorpha linza, which did not exhibit bands in the visible region of the optical absorption spectrum. However, since azide and cyanide (known inhibitors of heme enzymes) strongly inhibited this iodoperoxidase it was still suggested to be a heme enzyme. In 1983 Vilter (1983a, 1983b) investigated the presence of peroxidases in brown seaweeds (Phaeophyceae) and a non-heme haloperoxidase in the brown alga Ascophyllum nodosum was detected, which was insensitive towards inhibition by cyanide and lacked the Soret band in the absorption spectrum. The catalytic activity of this enzyme was found to be affected by the composition of the buffer solution. This observation led to the discovery that the bromoperoxidase could be inactivated by dialysis against EDTA in a citrate/phosphate buffer of pH 3.8. Incubation with vanadium(V) led to reactivation of the enzyme, whereas other metal ions were ineffective in the reconstitution of this enzyme (Vilter, 1984). These results were confirmed (Wever et al., 1985a, 1985b; De Boer et al., 1986a) and the presence of vanadium, essential for enzymatic activity, in these enzymes was established by these workers.

Recently, some other non-heme haloperoxidases have been isolated. In the bacterial bromoperoxidases present in Streptomyces aureofaciens (Krenn et al., 1988; Van Pée et al., 1987) and Streptomyces griseus (Zeiner, 1988) neither a heme group nor vanadium could be detected. The nature of the active site in these enzymes is still unknown. Also the fungal chloroperoxidase from Curvularia inaequalis (Liu et al., 1987) and the bacterial chloroperoxidase from Pseudomonas pyrrocinia (Wiesner et al., 1988) were devoid of heme. In these enzyme preparations both zinc and iron were detected, however, it is uncertain whether these metals are involved in the catalytic activity.

Distribution

Vanadium bromoperoxidases have been detected in and isolated from a number of brown algae (*Phaeophytae*), red algae (*Rhodophytae*) and from a lichen, which is a symbiosis of an alga and a fungus. To date, these include the enzymes from the brown seaweeds, *A. nodosum, Laminaria saccharina, Chorda filum* (De Boer *et al.*,1986a, 1986b), *Fucus distichus* and *Alaria esculenta* (Wever *et al.*, 1989). Vanadium bromoperoxidases have been characterized from the red seaweeds *Ceramium rubrum* (Krenn *et al.*, 1987), *Corallina pilulifera* (Krenn *et al.*, 1989) and *Corallina officinalis* (Yu and Whittaker, 1989). The enzyme from *C. pilulifera* was originally considered to be a non-heme iron containing peroxidase (Itoh, 1985, 1986, 1987). However, it has lately been demonstrated that the enzymes from both *Corallina* species contained vanadium at the active site which is essential for catalytic activity. A terrestrial organism, the lichen *Xanthoria parietina* which grows on rocks and stones also contains a vanadium bromoperoxidase (Plat *et al.*, 1987). Bromoperoxidases may be widespread in the marine environment; for example Hewson and Hager (1980) tested 72 algae of which 55 did contain bromoperoxidase activity. However, whether all these enzymes are vanadium bromoperoxidases is not yet known. Vanadium dependent iodoperoxidases have also been detected in the brown seaweeds *Fucus spiralis, F. serratis, F. vesiculosis* and *Pelvetia canaliculata* (De Boer, 1986b). These enzymes have not been characterized. Green seaweeds (*Chlorophytae*) also contain peroxidases (Hewson and Hager, 1980; Manthey and Hager, 1989) but the existence of a vanadium bromoperoxidase in green seaweeds has not been reported yet. Vanadium is also an essential element for growth of marine green algae and the brown macroalga *F. spiralis* (Fries, 1982) and unicellular green algae (Meisch and Bielig, 1975). Vanadium deficiency was found to cause reduced growth and impaired reproduction (Hopkins and Mohr, 1974). As may be expected, seaweeds also contain vanadium. Values of 0.3-10 μg per gram dry weight have been reported (Yamamoto *et al.*, 1970)

Molecular and Structural Aspects

Data on the molecular mass of the vanadium bromoperoxidases are scarce. On basis of gel-exclusion high-performance liquid-chromatography De Boer *et al.* (1986a) estimated a molecular mass of about 90 kDa for the enzyme from *A. nodosum* and 108 kDa for the enzyme from *L. saccharina* (De Boer *et al.*, 1986b). Titration experiments in which vanadate was added to the apo-enzyme from *A. nodosum* also suggest a molecular mass of 90 kDa (De Boer *et al.*, 1988a) which is in good agreement with recent sedimentation equilibrium studies (Tromp, *et al.*, 1989) in which a value of 97 kDa was found. SDS-polyacrylamide gel electrophoresis under denaturating conditions (that is boiling in the presence of 1% SDS and β-mercaptoethanol) yields for all vanadium bromoperoxidases known, a single subunit with a molecular mass of about 65 kDa (Table I). The discrepancy

Table I. Structural characteristics of vanadium bromoperoxidases.

Source	Molecular mass subunits (kDa)'	Isoelectric point	thermo-stability	Ref.
X. parietina	65	4.5	+ +	Plat et al., 1987
C. rubrum	58	3.9	−	Krenn et al., 1987
C. pilulifera	64	3.0	−	Itoh et al., 1987a
L. saccharina	66, 64	4.2	+	De Boer et al., 1986
A. nodosum I	67	5.0	+	ever et al., 1987a
A. nodosum II	70	5.0	+	Krenn et al., 1989a
C. officinalis	64	nd	nd	Yu & Whittaker., 1989

' The molecular mass was determined by SDS polyacrylamide gelelectrophoresis under denaturing conditions.

between molecular mass of the native enzyme from *A. nodosum* and that of the subunit suggests some abnormality in SDS binding. Alternatively, considering the rather unusual stability of some of these enzymes, it may be that these enzymes are defolded only incompletely under these conditions. The amino acid compositions of the various enzymes are in reasonably close agreement, with a predominance of glutamic and aspartic acid in line with the isoelectric points of these enzymes. (Table I). Some of the vanadium bromoperoxidases show immunological relationship, the vanadium bromoperoxidase from the red seaweed *C. rubrum* was partially immunologically identical to the vanadium enzyme from *A. nodosum* (Krenn et al., 1989a). It was also shown that the bromoperoxidases from different *Coralline* algae were immunologically related (Itoh et al., 1987a).

Some vanadium bromoperoxidases from temperate brown algae (Wever et al., 1985b) were quite thermostable whereas enzymes from the red seaweeds *C. rubrum* (Krenn et al., 1987) and *C. pilulifera* (Itoh et al., 1986) were less stable (see Table I). This phenomenon may be related to the habitat of these plants. Each algal species occupies a characteristic horizontal band or zone across the rock or shore surface in response to prevailing environmental conditions. Brown algae grow in the intertidal zone, so at low tide level these algae fall dry and are exposed to considerable heat in summer. In contrast, red algae always remain submerged even at low tide. The vanadium enzyme from the lichen *X. parietina*, which grows on sun-exposed rocks, is even more thermostable (Plat et al., 1987) than the peroxidases from the brown algae. However, the thermostability of enzymes from thermophilic bacteria is still considerably higher (Tombs, 1985). Some of the vanadium bromoperoxidases also exhibit a remarkable chemical stability. The enzyme from *A. nodosum* remained fully active in media containing appreciable amounts (up to 60% v/v) of alcohols (De Boer et al., 1987a). The vanadium bromoperoxidases from *A. nodosum* and *X. parietina* were also resistant towards storage in organic solvents such as methanol, ethanol or propanol (Wever et al., 1985b, 1987a; De Boer et al., 1987b). This bromoperoxidase from *A. nodosum* is also stable under turnover conditions. In an enzyme reactor in the presence of the substrates bromide, hydrogen peroxide and phenol red as a scavanger for HOBr, the enzyme remained active for 3 weeks (De Boer et al., 1987b). Even after sodium dodecylsulphate polyacrylamide gel electrophoresis the vanadium bromoperoxidases could be stained for enzymatic activity (Wever et al., 1985b; Plat et al., 1987; Krenn et al., 1987). Thus, if the enzyme samples are not boiled, the detergent SDS is unable to disrupt the protein conformation of these enzymes.

STRUCTURE OF THE ACTIVE SITE

One of the remarkable properties of the prosthetic group in vanadium bromoperoxidases is that it can be removed at low pH by dialysis against a citrate-phosphate buffer containing EDTA, which renders the enzyme inactive (Vilter, 1984; Wever et al., 1985a; De Boer et al., 1986a). Incubation of the enzyme with vanadate (VO_4^{3-}) at neutral pH results in recovery of the brominating activity. Reconstitution of enzymatic activity was inhibited by the structural analogues phosphate (PO_4^{3-}), arsenate (AsO_4^{3-}) and molybdate (MoO_4^{2-}) in a competitive manner with K_i values of 60, 120 and 98 μM, respectively (De Boer et al., 1986a, 1988a; Tromp et al., 1989). Further, the presence of phosphate stimulates the process of inactivation (Wever et al., 1987b). These observations suggest that the active site in native bromoperoxidase consists of a vanadate type of structure and that the valence state of the metal is 5 + (3d°). Further, the activation by vanadate suggests that at least 4 oxygen atoms are present as donors for the central vanadium(V)ion.

The content of vanadium in most bromoperoxidases is found to be less than stoichiometric (De Boer et al., 1986a, 1986b; Plat et al., 1987; De Boer et al., 1988a) but preparations can be reactivated by the addition of vanadate, and a correlation is observed between vanadium content and specific activity (De Boer et al., 1986; Yu and Whittaker, 1989). The bromoperoxidases from A. nodosum (De Boer et al., 1988a), C. officinalis (Yu and Whittaker, 1989) contain after reconstitution 1 V/mole and that from L. saccharina 2 V/mole (De Boer et al., 1986b). Apparently, vanadium is lost during purification or alternatively the bromoperoxidase in the seaweed is partly present in an apoform. This may well be the case since the concentration of vanadate in seawater is about 50 nM (Biggs and Swinehart, 1976) and the value of the dissociation constant for the equilibrium between apo-enzyme and vanadate was reported to be 35-55 nM at pH 8.5 and to increase rapidly at lower pH values (Vilter, 1984; Tromp et al., 1989). These enzymes have been studied by a variety of physiochemical methods and as will be discussed below; rather detailed structural data for the active site are available.

EPR studies have shown that a vanadyl type of spectrum is observed for the bromoperoxidases after reduction and from this observation it was concluded that the native bromoperoxidase contains vanadium in the 5 + oxidation state (De Boer et al., 1986b, 1988a). Very similar EPR parameters are obtained for enzymes from various species (Wever et al., 1987b, Krenn et al., 1989a) which indicate that the structure and the way the vanadyl group is ligated into the protein are very similar. It is possible to estimate the average ligand environment of VO^{2+} compounds by comparison of the g_o and A_o or $g_{\parallel}$ and $A_{\parallel}$ parameters with those of vanadyl chelates with known equatorial ligands (Chasteen, 1981; Boucher et al., 1969; Sakurai et al., 1988; Dutton et al., 1988). Comparison of the data for the bromo-peroxidases suggests a ligand environment consisting largely of oxygen and/or nitrogen atoms (De Boer et al., 1988a). Some evidence that nitrogen is present in the equatorial plane of the vanadyl cation of reduced bromoperoxidase was obtained from ESEEM spectra of bromoperoxidase which showed a pattern of ^{14}N frequencies, also found in model complexes which contained nitrogen in the equatorial plane of the vanadyl cation.

The oxovanadium IV site in reduced bromoperoxidase is also accessible for H_2O. Both $H_2^{17}O$ and D_2O affect the hyperfine line width in the EPR spectra of reduced bromoperoxidase (De Boer et al., 1988a). Further, the ESEEM spectra exhibit an intense [1]H-modulation which is completely replaced by a deuterium modulation when bromoperoxidase is dissolved in D_2O, instead of H_2O. NMR spectra of native vanadium bromoperoxidase show an unusual [51]V chemical shift and it was suggested (Vilter and Rehder, 1987; Rehder et al., 1988) that 6 or 7 highly electronegative oxygen functions are present.

Two research groups (Hormes et al., 1988; Arber et al., 1988, 1989) have reported K-edge X-ray absorption spectra for the enzyme in the oxidized state, in the presence of H_2O_2 and in the reduced state. According to Wong et al. (1984) the energy position of the pre-edge feature (1s-3d°) can be correlated with the coordination charge of vanadium complexes. Both groups subsequently used the position of this feature as an oxidation state marker and it was concluded that the vanadium atom in the native enzyme has an oxidation state of 5[+] which upon reduction is converted into the 4[+] oxidation state in line with the EPR data. However, recently a detailed X-ray absorption spectroscopy study was carried out of 27 different coordination compounds of vanadium in the 5[+] and 4[+] oxidation state and with various coordination numbers (Weidemann et al., 1989). No correlation was observed between the energy position of the pre-edge feature, the ligand electronegativities or the oxidation number of vanadium. Thus, the position of this marker and its intensity cannot be used to assign a coordination charge or size of the coordination charge of vanadium in bromoperoxidase.

EXAFS spectra for the native enzyme and the reduced enzyme were also obtained (Arber et al., 1989). The EXAFS and the XANES spectra of the two samples differ considerably and this suggests that upon reduction a significant change in the coordination environment takes place. On the basis of the spectra for the reduced enzyme a distorted octahedral geometry with two imidazoles at 2.11 Å, one oxygen at a short distance of 1.63 Å and three oxygens at 1.91 Å was proposed. Light atom backscattering was observed at 3 Å and 4.3 Å in both the native and reduced enzyme and this was suggested to be due to imidazole in the equatorial position. The presence of nitrogen is in line with the ESEEM studies (De Boer et al., 1988b). The EXAFS data for native bromoperoxidase are consistent with one oxygen at 1.61 Å, three oxygen atoms at 1.72 Å and 2 nitrogen atoms at 2.11 Å. The short vanadium oxygen distances are typical for those seen in vanadate or vanadium(V)-alkoxy systems (Scheidt, 1973; Caughlan et al., 1966). Whether indeed in native bromoperoxidase 2 nitrogen atoms are present or alternatively oxygen atoms cannot be decided yet. However, the presence of oxygen functions only would be consistent with the ^{51}V NMR experiments on native enzyme (Vilter and Rehder, 1987; Rehder et al., 1988). Figure 1 gives a proposal for the structure of the active site. Recently, the bromoperoxidase was crystallized (Müller-Fahrnow et al., 1988) and this is an important step in the elucidation of the tertiary structure. Future studies will show whether the proposed structure for the active site derived from all the data is indeed correct.

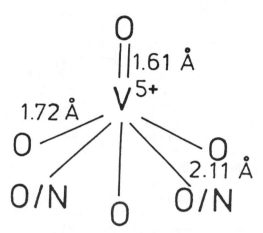

Fig. 1. Structure of the active center in vanadium bromoperoxidase as proposed by Arber et al. (1989).

It is not yet clear, how and which amino acid residues participate directly in the coordination of vanadate. Vanadate is known to form esters with hydroxyl groups (Gresser et al., 1986a, 1986b; Tracey et al., 1988; Tracey and Gresser, 1988), and to form complexes with amino groups of amino acid (Nechay et al., 1986; Rehder, 1988) and with carboxylates (Tracey et al., 1987). It is likely that one or more of these groups are involved in binding of vanadate. It is not very likely that a tyrosine is involved in the coordination to the vanadium(V) in bromo-peroxidase. Intense ligand to metal charge transfer bands in the visible spectrum such as those seen in mono-oxovanadium(V) phenolates (Cooper et al., 1982; Bonadies et al., 1986a, 1986b, 1987) have not been observed in bromoperoxidase (Vilter, 1983b, 1984; Wever et al., 1985b; Tromp et al., 1989). As pointed out, the EXAFS data suggest the presence of an imidazole. Steady-state kinetic studies (De Boer and Wever, 1988) have shown that a functional group with a pK_a of 5.7 is controlling the binding of H_2O_2 and this was ascribed to a histidine residue. Similarly, from changes in the EPR spectra of the reduced enzyme a group with a pK_a of about 5.4 was identified which was ascribed to a histidine or aspartate/glutamate residue. Alternatively, it is conceivable that this is due to ionization of a coordinated water molecule (De Boer et al., 1988a).

MECHANISM OF ACTION AND KINETIC PROPERTIES

An important question is of course, is there a redox shuttle operative during catalysis and do redox changes occur in the metal ion? The EPR results show that native bromoperoxidase contains vanadium in the 5^+ oxidation state, only upon reduction is the vanadyl type of EPR spectrum observed (De Boer et al., 1986a, 1986b; De Boer, 1988a; Yu and Whittaker, 1989). Addition of bromide or H_2O_2 or a combination of both did not result in the formation of the vanadyl state (Wever et al., 1987a; De Boer et al., 1988a). Further, when reduced bromo-peroxidase was treated with H_2O_2 or a combination of bromide and H_2O_2 no rapid reoxidation of the 4^+ oxidation state occurred. These findings do not favor a mechanism in which vanadium 5^+ is reduced to the 4^+ or 3^+ state and reoxidized to the 5^+ by H_2O_2. Rather a mechanism is operative in which H_2O_2 binds to the vanadium(V) which becomes activated and the complex then formed reacts with bromide to yield hypobromous acid. This proposal is supported by a number of observations. Steady-state kinetic studies (De Boer and Wever, 1988) which will be discussed later, point to a mechanism in which H_2O_2 first has to react with the enzyme before bromide is bound. In fact, if bromide reacts first with the enzyme it acts as an inhibitor. Further, the oxidizing properties of vanadium(V) peroxo-complexes are well documented (Secco, 1980; Mimoun et al., 1983; Mimoun et al., 1986) and inorganic vanadium(V) compounds with a 3d° con-figuration (Djordjevic, 1983) will easily form complexes with hydrogen peroxide. Finally, the formation of a stable bromoperoxidase peroxo-intermediate has re-cently been demonstrated. This intermediate rapidly decays when bromide is present (Tromp et al., 1989).

All vanadium bromoperoxidases studied do show pH optima (Table III) ranging from pH 5.5 to 7.2, the position of which is a function of the H_2O_2 and Br^- concentration (Wever et al., 1985b; De Boer and Wever, 1988). The K_m for bromide of the enzymes from seaweed varies from 1 to 15 mM which is in the same order of magnitude as the concentration of bromide (1 mM) in seawater (Riley and Chester, 1971). The enzyme from the lichen X. parietina is a notable exception, the high affinity for bromide may be a reflection of the habitat of this terrestrial organism (rocks and stones).

The specific activity of these enzymes varies considerably (Table III). However, most of these enzymes do contain vanadium at less than stoichiometric amounts and thus, relatively low specific activities are observed. After reconstitution, the bromoperoxidase from A. nodosum had a specific activity of 126 μ moles mono-chlorodimedone brominated per min per mg of protein and that from L. saccharina an activity of 615 units per mg of protein, a value which is similar to that (700 units/mg protein) reported for the enzyme from the red seaweed C. officinalis (Yu and Whittaker, 1989). However, even after reconstitution the specific activity of bromoperoxidase from C. pilulifera is considerably less (Krenn et al., 1989b). It should be noted that some of these enzymes easily aggregate under conditions of the reconstitution experiments and that this will hamper binding of vanadate (De Boer et al., 1986b) and prevent reactivation.

Table II. The amino acid composition of several vanadium bromoperoxidases.

Amino acid	Molar percentage				
	X. parietina[a]	C. pilulifera[b]	L. saccharina[c]	A. nodosum I[c]	A. nodosum II[d]
Asx	9.8	11.4	12.9	14.0	9.5
Thr	8.4	4.9	6.4	6.9	5.9
Ser	7.3	6.7	5.8	5.0	6.5
Glx	9.3	10.2	9.1	10.2	12.4
Pro	5.2	5.3	5.2	4.9	5.2
Gly	9.3	7.8	8.9	9.8	9.7
Ala	12.7	10.0	8.0	11.1	9.1
Val	6.1	6.5	5.5	6.2	6.2
Met	2.5	1.4	2.0	1.0	2.4
Ile	3.9	5.9	2.7	4.0	4.0
Leu	8.5	9.2	9.4	8.5	8.5
Tyr	3.4	2.2	3.1	1.7	3.1
Phe	5.8	5.7	3.5	5.7	7.0
Lys	2.0	4.3	3.7	1.6	1.7
His	1.3	1.4	1.7	1.2	1.0
Arg	3.0	5.1	4.7	4.0	4.0
Cys	nd	0.8	5.5	1.7	nd
Trp	nd	0.8	nd	nd	nd
Polarity[1] %	41.1	44.0	44.3	42.9	41.0

[1] The polarity was taken to be the sum of the molar percent values for Asx, Thr, Ser, Glx, Lys, His and Arg.
[a] Krenn et al. (1989a).
[b] Itoh et al. (1986).
[c] Wever et al. (1987).
[d] Krenn et al. (1989b).

Table III. Kinetic properties of vanadium bromoperoxidases.

Source	pH optimum	K_m for H_2O_2 (μM)	K_m for Br^- (mM)	Specific ctivity U/mg*	Ref.
X. parietina	5.5	870	0.03	79	Plat et al., 1987
C. rubrum	7.4	17	2	6.5	Krenn et al., 1987
C. pilulifera	6.0	92	11	16	Itoh et al., 1986
L. saccharina	6.5	27	1	134	De Boer et al., 1986a
A. nodosum I	6.0	27	14.8	60	Wever et al., 1987a
A. nodosum II	7.2	27	6.7	51	Krenn et al., 1989b

* The specific activity was measured for the enzymes as isolated.

An extensive steady-state kinetic analysis has been carried out for the enzyme from *A. nodosum* (De Boer and Wever, 1989) and the results show that hydrogen peroxide and bromide (as substrates) react with the enzyme in an ordered mechanism. Further, it was demonstrated that a reaction may occur between native bromoperoxidase and bromide to yield an inhibitory complex which is competitive with the binding of hydrogen peroxide. From this, it was concluded that hydrogen peroxide reacts first with the enzyme and that bromide is the second substrate in the catalytic cycle to yield a bromoperoxidase-HOBr species. No interaction could be demonstrated of this species with nucleophilic acceptors as 2-chlorodimedone, 5-phenylbarbituric acid and trans-4-hydroxy-cinnamic acid and it was concluded that the primary reaction products were the oxidized bromine species (Br_3^-, Br_2 and/or HOBr). To demonstrate the formation of these oxidized bromine species a stopped-flow/rapid scan spectrophotometer had to be used, since these oxidized bromine species will react extremely fast with H_2O_2 at neutral pH (De Boer and Wever, 1989; Kanofski, 1984, 1986). Similarly, Franssen *et al.* (1988) concluded from a study of the enzymatic bromination of barbituric acid and its derivatives that the enzyme releases free hypobromous acid which can either brominate the organic halogen acceptor or produce singlet oxygen by a competing reaction with hydrogen peroxide. Figure 2 gives the reaction scheme of the bromoperoxidase-induced bromination of organic compounds and the competitive reaction of HOBr with H_2O_2. Itoh *et al.* (1987, 1988) have also studied

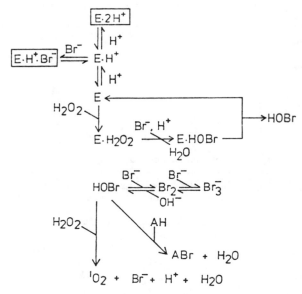

Fig. 2. Mechanism of action of vanadium bromoperoxidases and competitive reaction of HOBr with H_2O_2 or AH. AH is the organic substrate and ABr the brominated product. The enzyme species in boxes represent inhibited forms of bromoperoxidase.

the substrate specificity and stereospecificity of the halogenation reactions catalyzed by the bromoperoxidase of *C. pilulifera*. In line with the results of the Dutch group they now conclude in contrast to their original proposal (Itoh *et al.*, 1987c) that the enzyme had no specific site for the organic substrates studied, and that regio or stereospecificity is absent due to formation of HOBr by bromoperoxidase. If an enzyme is able to produce such oxidizing bromine species in solution, this requires that the bromoperoxidase is not inactivated by its own product. Resistance towards 0.5 mM HOBr was indeed observed (De Boer *et al.*, 1987a). Interestingly, Sauvageau (1926) detected free bromine production by some seaweeds.

De Boer and Wever (1988) showed that inhibition of enzymatic activity at low pH was due to protonation of an ionizable group in the enzyme. Since hydrogen peroxide was unable to bind to the native enzyme when this group was protonated, this functional group is probably responsible for the pH-controlled binding of H_2O_2 (second-order rate constant $2.5 \cdot 10^6 M^{-1} \cdot s^{-1}$ at pH <6). The pK_a for this group was 5.7 and is ascribed to a histidine residue. From EPR experiments it also appeared that an acid/base group was present near the vanadium(IV) ion with a pK_a of 5.4. High concentrations of H_2O_2 did not damage the enzyme and inhibition by H_2O_2 was not observed (Wever *et al.*, 1985; De Boer and Wever, 1988). Bromide reacts in a complex fashion with the enzyme-hydrogen peroxide complex. At least two pH-dependent intermediates are present which react with bromide (second-order rate constant $1.7 \ 10^5 \ M^{-1} \cdot s^{-1}$) to yield the oxidized bromine species as the enzymatic product. The presence of excess bromide also leads to inhibition of the bromoperoxidase from *A. nodosum*. The value of the K_i for bromide is a function of the pH and varies from 228 mM to more than 1 M. No significant inhibition is observed for inhibitors as cyanide and azide in the mM concentration (Wever *et al.*, 1985b; De Boer *et al.*, 1986a,1986b) and also 0.5 M chloride had no effect on the enzymatic activity. In contrast, the bromoperoxidase from the lichen *X. parietina* is inhibited significantly by 50 mM F^-, Cl^- or Br^- (Plat *et al.*, 1987) and 1-5 mM nitrate (Wever *et al.*, 1987a). This enzyme also has a much higher affinity for bromide than the enzymes isolated from seaweed (Table III).

Physiological Aspects, Formation of Halometabolites and Impact on the Biosphere

Marine organisms are a rich source of a wide variety of halometabolites and several reviews concerning the nature and biological activity have been published (Faulkner, 1984; Fenical, 1975, 1979, 1982; McConnel and Fenical, 1977b). The majority of halogenated compounds isolated from the marine environment are bromometabolites. For example, a highly brominated antibiotic (five bromine atoms per molecule) was isolated from the marine bacterium *Pseudomonas bromoutilis* (Burkholder *et al.*, 1966). The marine worm *Balanoglossus biminiensis* produces large amounts of 2,6 dibromophenol as a defence mechanism to ward off predators (Ashworth and Corrmier, 1967). The presence of bromophenols in Rhodomelaceae and other algae has been reported (Pederson and Da Silva, 1973; Pederson et al., 1974; Pederson and Fries, 1975; Manley and Chapman, 1978) and these compounds were shown to be antibacterial (King, 1986) and antifungal (McLachlan and Craigie, 1966). Furthermore, halometabolites are also found as intermediates in the biosynthesis of complex halogenated compounds. Strong presumptive evidence has been accumulated that in the marine environment brominated intermediates may be involved in the cyclization of isoprenoids to yield terpenes and in rearrangement reactions of linear ketones and terpenoid molecules (Fenical, 1982).

Also simple volatile halohydrocarbons are produced. The tropical red seaweed Asparagopsis taxiformis for example, produces large amounts (3-5% dry weight) of halomethane derivatives with bromoform as the major volatile metabolite (Burreson et al., 1976; McConnel and Fenical, 1977; Fenical, 1974). These halogenated methanes ($CHBr_3$, $CHBr_2Cl$, $CHCl_3$ and CCl_4) which are potent biocides are apparently formed by the classical haloform reaction of methyl or carboxy methyl ketones (Burreson et al., 1976; Moore, 1977). Members of the family Bonnemaisoniaceae have been shown to produce a variety of halogenated ketones (Siuda et al., 1975; McConnel and Fenical, 1977). Evidence that these halometabolites and the volatile bromomethanes are the products of enzyme-catalyzed bromination of ketones was presented by Theiler et al. (1978). They showed that a bromoperoxidase extracted from the marine red alga Bonnemaisonia hamifera was capable to incorporate bromine into a number of organic substrates, leading to formation of volatile brominated hydrocarbons, like dibromomethane, bromoform and 1-pentylbromide.

Methylchloride is also produced in considerable amounts by the sea. Lovelock and coworkers (1973, 1975) who measured aerial and seawater concentrations of halohydrocarbons demonstrated that this halogenated compound does not originate from industrial polution, but arises from biological activity in seawater. The concentration of halohydrocarbons was highest in and around tropical ocean waters and was especially abundant over algae beds (Lovelock et al., 1973; Lovelock, 1975). The annual global emission rate required to maintain the measured steady-state level of methylchloride was estimated to be 2×10^7 tons per year (Hager, 1982) This instable compound is thought to be synthesized by marine algae, excreted into the ocean water and diffusing into the upper atmosphere. Whether also a haloperoxidase is involved in the biosynthesis of this compound is not yet clear. It is interesting to note that methylchloride may actually be the product of a nucleophilic reaction between methyliodide and chloride ions (Zafiriou, 1975) since methyliode is a known product of marine algae (Lovelock, 1975; Burreson et al., 1976).

Recently, Gschwend et al. (1985) observed that brown, red and green algae release large quantities of volatile brominated methanes like bromoform, dibromochloromethane and dibromomethane into the marine ecosystem. For A. nodosum a mean rate of release of $CHBr_3$ of 4.5 microgram per gram of dry algae per day was reported. It is very likely that the vanadium bromoperoxidases present

(Krenn *et al.*, 1989b) in this seaweed are involved in the release of this and other compounds. It is of interest to note that one of these enzymes appears to be located outside on the thallus surface and can be considered to be extracellular (Krenn *et al.*, 1989b). This is in line with the suggestions by Wolk (1968) and Gschwend *et al.* (1985) that brominated compounds are synthesized near the plant surface. Gschwend *et al.*, also calculated the rate of release of these compounds by brown, red and green algae and they concluded that these macroalgae are an important source of bromine-containing material released to the atmosphere. Assuming a release of 1-10 μg Br in the form of organobromides per gram per day and a global algal biomass of 10^{13} g algae (Waaland, 1981) it is possible to estimate an annual global input of 10^4 ton per year. This is of the same order of magnitude as the reported anthropogenic input of organobromides.

Interestingly, high levels of brominated organic species of which bromoform was the main component have been detected in the Arctic atmosphere (Berg *et al.*, 1984). Further, the bromine content of Arctic aerosol at Point Barrow (Alaska) and Alert (Canadian High Arctic) shows an annual sharp maximum between February and May (Berg *et al.*, 1983; Sturges and Barrie, 1988) which was also observed in the aerosol over Spitsbergen at the same time (Berg *et al.*, 1983). When compared to the natural background levels their results show that the bromine concentrations in the Arctic troposphere are extraordinarily high, the highest found anywhere in the world. Direct measurements of atmospheric methylbromide and bromoform at Barrow also showed a large seasonal variation with maximal bromoform concentrations in winter and minimal in summer (Cicerone *et al.*, 1988). There is substantial evidence that the source of bromine containing particles and bromoform is biogenic and is produced in the oceans in the northern part of the northern hemisphere. Dyrssen and Fogelqvist (1981) measured the bromoform concentration in the Arctic Ocean near Spitsbergen. They showed that bromoform is present at all depths in the eastern Arctic Ocean and the profiles of depths versus concentration showed that in-shore algae belts produce bromoform in substantial amounts. Similar data were obtained for the Skagerrak and Idefjorden (Fogelqvist and Krysel, 1986). Macroalgae are probably not the only source of bromoform, the data of Fogelqvist (1985) also indicate that plankton may contribute to the formation of bromoform in oceans.

There is considerable interest in this phenomenon since Barrie *et al.* (1988) showed that at polar sunrise ozone destruction occurs in the lower Arctic atmosphere at Alert and that this strongly correlated with the presence of filterable bromine in aerosols. Organobromides like bromoform are precursors of filterable bromine in aerosols and as has been shown both bromoform production and ozone destruction show a seasonal cycle. It is very likely that the vanadium bromoperoxidases present in seaweeds in the Arctic Ocean are responsible for the production of bromoform and other halogenated compounds and it has been proposed (Wever, 1988) that this enzymatic activity is linked to the observed ozone destruction at ground level in the Arctic. This proposal follows the suggestion made by Theiler *et al.* (1978) that volatile halohydrocarbons in the biosphere result from peroxidase-catalyzed halogenation reactions. If so, these vanadium enzymes in seaweeds and their products have had a large impact on the chemistry and composition of the atmosphere on earth during evolution.

Acknowledgements

We wish to thank Ms. M. van der Kaaden for her help in the literature search and Mr. M.L. Dutrieux for preparing the manuscript. This work is supported in part by the Netherlands Foundation for Chemical Research (S.O.N.) with financial aid from the Netherlands Organisation for Scientific Research (N.W.O.).

References

Arber, J.M., De Boer, E., Eady, R.R., Garner, C.D., Hasnain, S.S., Smith, B.E. and Wever, R. (1988) The second Intern. Conf. on Biophysics and Synchrotron radiation, 4–8 July,Chester, U.K. p.72.

Arber, J.M., de Boer, E., Garner, C.D., Hasnain, S.S. and Wever, R. (1989) Biochemistry 28, 7968–7973.

Ashworth, R.B. and Cormier, M.J. (1967) Science 155, 1558–1559.

Atkins, W.R.G. (1914) Scient. Proc. Roy. Dublin Soc. 14, 199–206.

Bach, A.N. and Chodat, R. (1903) Ber. Dtsch. Chem. Ges. 36, 600–605.

Bakkenist, A.R.J., de Boer, J.E.G. and Wever, R. (1980) Biochim. Biophys. Acta 613, 337–348.

Barrie, L.A., Bottenheim, J.W., Schnell, R.C., Crutzen, P.J. and Rasmussen, R.A. (1988) Nature 334, 138–141.

Battelli, F. and Stern, L. (1908) Biochem. Z. 13, 44–88.

Berg, W.W., Sperry, P.D., Rahn, K.A. and Gladney, E.S. (1983) J. Geophys. Res. 88, 6719–6736.

Berg, W.W., Heidt, L.E., Pollock, W., Sperry, P.D. and Cicerone, R.J. (1984) Geophys. Res. Lett. 11, 429–432.

Biggs, W.R. and Swinehart, J.H. (1976) in Metals Ions in Biological Systems, (Sigel, H., Ed.) Vol. 6, Chapter 2, Marcel Dekker Inc., New York.

Bonadies, J.A., Pecoraro, V.L. and Carrano, C.J. (1986) J. Chem. Soc. Chem. Commun. 1218–1219.

Bonadies, J.A., Butler, W.M., Pecoraro, V.L. and Carrano, C. (1987) J. Inorg. Chem 26, 1218–1222.

Boucher, L.J., Tyn, E.C. and Yen, T.F. (1969) in Electron Spin Resonance of Metal Complexes (Yen, T.F., Ed.) pp. 11–130, Plenum Press, New York.

Burkholder, P.R., Pfister, R.M. and Leitz, F.H. (1966) Appl. Microbiol. 14, 649–653.

Burreson, B.J., Moore, R.E. and Rohler, P. (1976) J. Agric. Food Chem. 24, 856–861.

Caughlan, C.N., Smith, H.M. and Waterpaugh, K. (1966) Inorg. Chem. 5, 2131–2134.

Chance, B. (1949) Arch. Biochim. Biophys. 22, 224–252.

Chasteen, N.D. (1981) in Biological Magnetic Resonance (Berliner, L. and Reuben, J., Eds.) 3, pp. 53–119, Plenum Press, New York.

Cicerone, R.J., Heidt, L.E. and Pollock, W.H. (1988) J. Geophys. Res. 93, 3745–3749.

Cooper, R.S., Bai Koh, Y. and Raymond, K.N. (1982) J. Am. Chem. Soc. 104, 5092–5102.

De Boer, E., Van Kooyk, Y., Tromp, M.G.M., Plat, H. and Wever, R. (1986a) Biochim. Biophys. Acta 869, 48–53.

De Boer, E., Tromp, M.G.M., Plat, H., Krenn, G.E. and Wever, R. (1986b) Biochim. Biophys. Acta 872, 104–115.

De Boer, E., Plat, H. and Wever, R. (1987a) in Biocatalysis in Organic Media (Laane, C., Tramper, J. and Lilly, M.D., Eds.) 29, 317–322 Elsevier, Amsterdam.

De Boer, E., Plat, H., Tromp, M.G.M., Wever, R., Franssen,M.C.R., Van Der Plas, H.C., Meijer, E.M. and Schoemaker, H.E. (1987b) Biotechnol. Bioeng. 30, 607–610.

De Boer, E., Boon, K. and Wever, R. (1988a) Biochemistry 27, 1629–1635.

De Boer, E., Keijzers, C.P., Klaasen, A.A.K., Reijerse, E.J., Collison, D., Garner, C.D. and Wever, R. (1988b) FEBS Lett. 235, 93–97.

De Boer, E. and Wever, R. (1988) J. Biol. Chem. 263, 12326–12332.

Djordjevic, C. (1982) Chem. in Britain 18, 554–557.

Dolin, M.J. (1957) J. Biol. Chem. 225, 557–573.

Dunford, H.B. and Stillman, J.S (1976) Coord. Chem. Rev. 19, 187–251.

Dutton, J.C., Fallon, G.D. and Murray, K.S. (1988) Inorg. Chem. 27, 34–38.

Dyrssen, D. and Fogelqvist, E. (1981) Oceanol. Acta 4, 313–317.

Faulkner, D.J. (1984) Natural Product Reports 1, 251–280.

Fenical, W. (1975) J. Phycol. 11, 245–259.

Fenical, W. (1979) Recent Adv. Phytochem. 13, 219–239.

Fenical, W. (1982) Science 215, 923–928.

Fogelqvist, E. (1985) J. Geophys. Res. 90, 9181–9193.

Fogelqvist, E. and Krysell, M. (1988) Mar. Poll. Bull. 17, 378–382.

Franssen, M.C.R., Jansma, J.D., Van Der Plas, H.C., De Boer, E. and Wever, R. (1988) Bioorg. Chem. 16, 352–356.

Fries, L. (1982) Planta 154, 393–396.

Gresser, M.J. and Tracey, A.S. (1986a) J. Am. Chem. Soc. 108, 1935–1939.

Gresser, M.J. and Tracey, A.S. (1986b) Proc. Natl. Acad. Sci. 83, 609–613.

Gschwend, P.M., MacFarlane, J.K. and Newman, K.A. (1985) Science 227, 1033–1035.

Hager, L.P., Morris, D.R., Brown, F.S. and Eberwein, H. (1966) J. Biol. Chem. 241, 1769–1777.

Hager, L.P. (1982) Basic Life Sciences 19, 415–429.

Harrison, J.E. and Schultz, J. (1978) Biochim. Biophys. Acta 536, 341–349.

Hewson, W.D. and Hager, L.P. (1979) in The Porphyrins (Dolpin, D., Ed.) Vol. 7, p.p. 295–332, Academic Press, New York.

Hewson, W.D. and Hager, L.P. (1980) J. Phycol. 16, 340–345.

Hopkins, Jr. L.L. and Mohr, H.E. (1974) Fed. Proc. Am. Soc. Exp. Biol. 33, 1773–1775.

Hormes, J., Kuetgens, U., Chauvistre, R., Schreiber, W., Anders, N., Vilter, H., Rehder, D. and Weidemann, C. (1988) Biochim. Biophys. Acta 956, 293–299.

Itoh, N., Izumi, Y. and Yamada, H. (1985) Biochem. Biophys. Res. Commun. 131, 428–435.

Itoh, N., Izumi, Y. and Yamada, H. (1986) J. Biol. Chem. 261, 5194–5200.

Itoh, N., Izumi, Y. and Yamada, H. (1987a) J. Biol. Chem. 262, 11982–11987.

Itoh, N., Hassan, A.K.M.O., Izumi, Y. and Yamada, H. (1987b) Biochemistry Int. 15, 27–33.

Itoh, N., Izumi, Y. and Yamada, H. (1987c) Biochemistry 26, 282–289.

Itoh, N., Hassan, A.K.M.O., Izumi, Y. and Yamada, H. (1988) Eur. J. Biochem. 172, 477–484.

Kanofsky, J.R. (1984) J. Biol. Chem. 259, 5596–5600.

Kanofsky, J.R. (1986) Biochem. Biophys. Res. Commun. 134, 777–782.

King, G.M. (1986) Nature 323, 257–259.

Krenn, B.E., Plat, H. and Wever, R. (1987) Biochim. Biophys. Acta 912, 287–291.

Krenn, B.E., Plat, H. and Wever, R. (1988) Biochim. Biophys. Acta 952, 255–260.

Krenn, B.E., Izumi, Y., Yamada, H. and Wever, R. (1989a) Biochim. Biophys. Acta 998, 63–68.

Krenn, B.E., Tromp, M.G.M. and Wever, R. (1989b) J. Biol. Chem. 264, 19207–19292.

Ladenstein,R. and Wendel, A. (1976) J. Mol. Biol. 104, 877–882.

Linossier, G. (1898) C.R. Soc. Biol., Paris 50, 373–375.

Liu, T.-N.E., M'Timkulu, T., Geigert, J., Wolf, B., Neidleman, S.L., Silva, D. and Hunter-Cevera, J.C. (1987) Biochem. Biophys. Res. Commun. 142, 329–333.

Lovelock, J.E., Maggs, R.J. and Wade, R.J. (1973) Nature 241, 194–196.

Lovelock, J.E. (1975) Nature 256, 193–194.

MacDonald, Jr., C.F. and Smith, H.C. (1909) J. All. Dent. Soc. 4, 346–352.

Manley, S.L. and Chapman, D.J. (1978) Febs Lett. 93, 97–101.

Manthey, J.A. and Hager, L.P. (1989) Biochem. 28, 3052–3057.

McConnell, O.J. and Fenical, W. (1977a) Phytochem. 16, 367–369.

McConnell, O.J and Fenical, W. (1977b) Tetrahedron Lett. 48, 4159–4162.

McLachlan, J. and Craigie, J.S. (1966) J. Phycol. 2, 133–135.

Meisch, H.-U. and Bielig, H.-J. (1975) Arch. Microbiol. 105, 77–82.
Morrison, M. and Schonbaum, G.R. (1976) Ann. Rev. Biochem. 45, 861–888.
Moore, R.E. (1977) Acc. Chem. Res. 10, 40–47.
Mimoun, H., Saussine, L., Daire, E., Postel, M., Fischer, J. and Weiss, R. (1983) J. Am. Chem. Soc. 105, 3101–3110.
Mimoun, H., Mignard, M., Brechelot, P. and Saussine, L. (1986) J. Am. Chem. Soc. 108, 3711–3718.
Müller-Fahrnow, A., Hinrichs, W., Saenger, W. and Vilter, H. (1988) FEBS Lett. 239, 292–294.
Murphy, M.J. and O'hEocha, C. (1973) Phytochem. 12, 61–65.
Nechay, B.R., Nanninga, L.B. and Nechay, P.S.E. (1986) 251, 128–138.
Neidleman, S.L. and Geigert, J. (1986) in Biohalogenation: Principles, Basic Roles and Applications, Ellis Horwood Ltd., Chichester.
Paul, K.G. (1963) in The Enzymes (Boyer, P.D., Lardy, H. and Myrback, K., Eds) Vol. 8, pp.227–274, Acadeemic Press, New York and London.
Pedersen, M. and Fries, L. (1975) Z. Pflanzenphysiol. 74, 272–274.
Pedersen, M. and DaSilva, E.J. (1973) Planta 115, 83–86.
Pedersen, M., Saenger, P. and Fries, L. (1974) Phytochem. 13, 2273–2279.
Petersson, S. (1940) Kungl. Fysiogr. Slsk. Lund Forhandl. 10, 171–182.
Planche (1810) Bull. Pharm. 2, 578–580.
Plat, H., Krenn, B.E. and Wever, R. (1987) Biochem. J. 248, 277–279.
Reed, G.B. (1915) Bot. Gaz. 59, 407–409.
Rehder, D., Weidemann, C., Duch, A. and Priebsch, W. (1988) Inorg. Chem. 27, 584–587.
Rehder, D. (1988) Inorg. Chem. 27, 4312–4316.
Riley, J.P. and Chester, R. (1971) in Introduction to Marine Chemistry, pp. 60–104, Academic Press, London and New York.
Rönnerstrand, S. (1946) Kungl. Fysiogr. Slsk. Lund Forhandl. 16, 117–130.
Sakurai, H., Hirata, J. and Michibita, H. (1988) Inorg. Chim. Acta 152, 177–180.
Saunders, B.C., Holmes-Siedle, A.G. and Stark, B.P. (1964) in Peroxidases. The properties and uses of a versatile enzyme and some related catalysts, Butterworth, London.
Sauvageau, C. (1926) Bull. Stat. Biol. Arc. 23, 5–23.
Scheidt, W.R. (1973) Inorg. Chem. 12, 1758–1761.
Schonbein, C.F. (1855) Verh. Naturf. Ges. Basel. 1, 339–355.
Secco, F. (1980) Inorg. Chem. 19, 2722–2725.
Siuda, J.F., Van Blaricom, G.R., Shaw, P.D., Johnson, R.D., White, R.H., Hager, L.P. and Rinehart, K.L. (1975) J. Am. Chem. Soc. 97, 937–938.
Sturges, W.T. and Barrie, L.A. (1988) Atmos. Env. 22, 1179–1194.
Tamiya, H. (1934) Planta 23, 284–288.
Theiler, R., Cook, J.C., Hager, L.P. and Siuda, J.F. (1978) Science 202, 1094–1096.
Tombs, M.P. (1985) J. Appl. Biochem. 7, 3–24.
Tracey, A.S., Gresser, M.J. and Galeffi, B. (1988) Inorg. Chem 27, 157–161.
Tracey, A.S. and Gresser, M.J. (1988) Inorg. Chem. 27, 2695–2702.
Tracey, A.S., Gresser, M.J. and Parkinson, K.M. (1987) Inorg. Chem. 26, 629–639.
Tromp, M.G.M., Olafsson, G., Krenn, B.E. and Wever, R. (1990) Biochem. Biophys. Acta, submitted.
Van Pée, K.-H., Sury, G. and Lingens, F. (1987) Biol. Chem. 368, 1225–1232.
Vilter, H. (1983a) Bot. Mar. 26, 429–435.
Vilter, H. (1983b)) Bot. Mar. 26, 451–455.
Vilter, H. (1984) Phytochem. 23, 1387–1390.
Vilter, H. and Rehder, D. (1987) Inorg. Chim. Acta 136, L7–L10.
Waaland, R.J. (1981) in The Biology of Seaweeds (Lobban, C.S. and Wynne, M.J., Eds.) pp. 726–741, University of California Press, Berkely and Los Angeles.
Weidemann, C., Rehder, D., Kuetgens, N., Hirmes, J. and Vilter, H. (1989) Chem. Physics, in press.
Wever, R., Plat, H. and De Boer, E. (1985a) Rev. Port. Quim. 27, 169–170.
Wever, R., Plat, H. and De Boer, E. (1985b) Biochim. Biophys. Acta 830, 181–186.
Wever, R. Krenn, B.E., Offenberg, H., De Boer, E. and Plat, H. (1987a) Progress in Clin. and Biol. Res. (King, T.E., Mason, H.S. and Morrison, M., Eds.) 274, 477–493, Liss Inc., New York.

Wever, R., De Boer, E., Krenn, B.E. and Plat, H. (1987b) Rec. Trav. Chim. Pays-Bas 106, 181.

Wever, R. (1988) Nature 335, 501.

Wever, R., Olafsson, G., Krenn, B.E. and Tromp, M.G.M. (1989) Abstr. 32nd IUPAC Stockholm, no. 210.

Wiesner, W., Van Pee, K.-H. and Lingens, F. (1988) J. Biol. Chem. 263, 13725–13732.

Wolk, C.P. (1968) Planta 78, 371–378.

Wong, J., Lytle, F.W., Messner, R.P. and Maylotte, D.H. (1984) Phys. Rev. 30, 5596–5610.

Yamamoto, T., Fujita, T. and Ishibashi, M. (1970) Rec. Oceanogr. Works Jpn. 10, 125–135.

Yu, H. and Whittaker, J.W. (1989) Biochem. Biophys. Res. Commun. 160, 87–92.

Zafiriou, O.C. (1975) J. Mar. Res. 33, 73–79.

Zeiner, R., Van Pée, K.-H. and Lingens, F. (1988) J. Gen. Microbiol. 134, 3141–3149.

VI. Vanadium Nitrogenases

ROBERT R. EADY
Nitrogen Fixation Unit, University of Sussex, Brighton, BN1 9RQ, U.K.

Introduction

Recent research has shown that vanadium has a role in biological nitrogen fixation. It is now clear that in *Azotobacter* species there are three genetically distinct but related nitrogenase systems. One system is based on Mo and has properties similar to Mo-nitrogenase of many other nitrogen fixing organisms, another is based on V and forms the focus of this chapter, and a third system which contains Fe and only low levels of Mo or V and is discussed only briefly here.

Which of these systems is expressed is dependent on the availability of Mo and V in the growth medium. In the presence of Mo the Mo-independent nitrogenases are not synthesised, a regulatory phenomenon which delayed their discovery by many years.

Biological nitrogen fixation, the conversion of atmospheric N_2 to a form utilizable by plants, is catalysed by the enzyme nitrogenase. Early studies on the effect of transition metals in stimulating the N_2-dependent growth of the soil bacterium Azotobacter showed both Mo (Bortels 1930) and V (Bortels 1936) to share this ability. However, subsequent biochemical and genetical research resulted in the view that Mo had an essential and unique role in nitrogenase function. It is only within the last 5 years that a role for vanadium in nitrogen fixation been unequivocally established.

The dominating role for molybdenum stemmed initially from the biochemical characterization of nitrogenase. Although this enzyme is found in relatively few groups of bacteria, biochemical studies on purified nitrogenase from organisms of quite different taxonomic groups showed the enzymes to be very similar. All purified nitrogenase could be separated into a molybdoprotein (MoFe protein) and an iron containing protein (Fe protein). The presence of Mo in the MoFe proteins in a molybdenum and iron cofactor centre (FeMoco) which is probably the site at which N_2 is reduced, rationalized the long-recognised physiological effect of Mo in stimulating the N_2-dependent growth of Azotobacter.

Parallel genetic studies showed that the structural genes encoding the Mo-nitrogenase proteins had been highly conserved in evolution, and were often clustered on the chromosome with genes involved in Mo uptake and FeMoco biosynthesis. Not surprisingly, it became generally accepted that N_2 fixation occurred by a single route involving Mo as an essential component of nitrogenase (See Eady 1986; Shah *et al.* 1984).

This unifying view was challenged in 1980 when it was proposed by Bishop and his colleagues that Azotobacter possessed a Mo-independent nitrogenase system.

N. Dennis Chasteen (ed.), Vanadium in Biological Systems, 99–127.
© 1990 *Kluwer Acadmic Publishers. Printed in the Netherlands.*

The basis for this proposal was the observation that some mutants of *Azotobacter vinelandii*, with lesions in the structural genes of nitrogenase and unable to grow on N_2, regained this ability provided that Mo was omitted from the growth medium (Bishop *et al.* 1980). Despite the accumulation of data supporting this hypothesis, (Bishop *et al.* 1982; Page and Collinson 1982; Premakumar *et al.* 1984), it was not generally accepted until the techniques of recombinant DNA technology were used to construct mutant strains of *A vinelandii* in which the genes encoding the structural polypeptides of Mo-nitrogenase were specifically removed (Bishop *et al.* 1986a). The demonstration that such strains could grow on N_2 provided that Mo was not added to the growth medium established the separate existence of Mo-independent nitrogenase in Azotobacter (Bishop *et al.* 1986b).

A comparable deletion strain of *Azotobacter chroococcum* was constructed (Robson 1986) and growth shown to be dependent on vanadium and inhibited by Mo. Vanadium was effective at low concentrations (1 to 10 nM) in stimulating both growth and nitrogenase activity of this strain which was the source of the first vanadium-nitrogenase to be isolated (Robson *et al.* 1986).

Mutant strains of *A. chroococcum* carrying deletions of both Mo and V nitrogenase structural genes are unable to grow on N_2 (Robson *et al.* 1989). In contrast, comparable double deletion strains of *A. vinelandii* can utilise N_2 for growth provided that both Mo and V are absent from the growth medium (Pau *et al.* 1989). *A. vinelandii* but not *A. chroococcum*, thus has a third nitrogenase system which when purified contains only low levels of Mo or V (Chisnell *et al.* 1988). The structural genes encoding the third nitrogenase have been cloned and sequenced and shown to be homologous with those of Mo and V nitrogenase systems (Joerger *et al.* 1989).

All three types of nitrogenase have similar requirements for activity – a low potential electron donor, MgATP and anoxic conditions. They can also be separated into two protein components, a distinct Fe protein which together with the MoFe protein, VFe protein or a component lacking significant amounts of Mo or V, make up the active enzyme.

Developments in our understanding of V-nitrogenase have been rapid because of the comparative nature of the interpretation of data and because many of the methods developed for the study of Mo-nitrogenase are directly applicable to this system. Consequently the background provided by the better characterized Mo-nitrogenase is discussed here using review references. An outline of the genetics of nitrogen fixation, so important in proving that V-nitrogenase existed, is also presented since it is relevant to both the structure of V-nitrogenase and the rationale for the assay of cofactor centres that nitrogenase contain.

Genetics of Nitrogen Fixation

KLEBSIELLA PNEUMONIAE – THE MODEL NITROGENASE SYSTEM

The genetics of nitrogen fixation is complex and is best understood in the enteric diazotroph *Klebsiella pneumoniae*, an organism which only has a Mo-nitrogenase system. The twenty *nif* genes (*nif* ≡ nitrogen fixation) in this organism have been sequenced and are clustered on a 24Kb region of the chromosome arranged in eight transcriptional units (See Merrick 1988). The function of *nif* gene products in *K. pneumoniae* was assigned initially on the basis of properties of extracts of wild-type and point as Mu insertion Nif⁻ mutants (See Eady 1986). It has been shown subsequently that many other diazotrophs have *nif* gene products with homologous properties.

The formation of an active Mo-nitrogenase requires, in addition to the genes encoding the subunits of the MoFe protein (*nifDK*) and the Fe protein (*nifH*), those for FeMoco biosynthesis (*nifHBENVQ*) and activation of the Fe protein polypeptide (*nifM*) (see Figure 1). The detailed biochemistry of FeMoco biosyn-

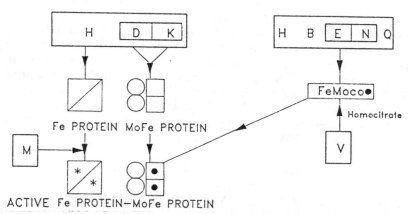

Fig. 1. Genes involved in the synthesis of active Mo-nitrogenase in *K pneumoniae*. The apo-Fe protein is activated by NIFM in some unknown way, but the transfer of *nifH* and *nifM* to the non-nitrogen fixing organism *E.coli* is sufficient for the synthesis of active Fe protein (Paul and Merrick 1989). Nothing is known about the genes involved in the biosynthesis of 'P' clusters of the MoFe proteins. Of the genes involved in FeMoco biosynthesis (for review see Hoover *et al.* 1988) the products NIFEN form an $\alpha_2 \beta_2$ complex and have Cys spacing analogous to the NIFDK polypeptides. NIFV appears to be a homocitrate synthase and NIFQ enables Mo to be utilized when present at low concentrations in the growth medium. To be effective in FeMoco biosynthesis NIFH does not have to be catalytically active in nitrogenase function.

thesis and post translational modification steps are not known, but *nifB*⁻ mutants of *K. pneumoniae* and *A. vinelandii* have been used as a source of MoFe protein lacking cofactor to provide a basis for the assay of isolated FeMoco. In addition to these genes, nitrogenase activity *in vivo* involves the function of the two *nif* specific genes involved in the generation of low potential reducing equivalents, *nifJ* encoding a pyruvate flavodoxin oxidoreductase, and *nifF*, encoding a flavodoxin, capable of donating electrons to the Fe protein. Other genes in this region, *nif*TYUSWZ have no function assigned to them.

Nitrogenase function is an energetically expensive process since the enzyme can constitute up to 10% of the soluble cell protein and at least 16 ATP molecules are hydrolysed for each N_2 molecule reduced. Nitrogenase synthesis is tightly regulated in response to the ambient dissolved oxygen tension and the availability of fixed-N to the organisms (See Dixon 1989). The enzyme is not synthesised when adequate fixed nitrogen is available to the organism or when the dioxygen tension would irreversibly inactivate this oxygen-sensitive enzyme (See Hill 1988).

Two *nif*-specific genes are involved in this regulatory process, *nifL* and *nifA*. Transcription from all *nif* promoters requires a minor form of the RNA polymerase sigma factor (σ^{54}, the *ntrA* (*rpoN*) gene product) for recognition by RNA polymerase, and all *nif* promoters other than that of the regulatory *nifLA* operon require the *nifA* product as an activator. Positive control of *nif* transcription by activator proteins involves DNA-protein interaction some distance upstream from the transcription initiation site. Negative control in the presence of repressive concentrations of dioxygen or fixed-N is achieved by the *nifL* product modifying the activator properties of the *nifA* product by some unknown mechanism. Expression of the *nifLA* operon itself in response to fixed-N levels, is controlled by the *NtrBC* cascade system in which the phosphorylated form of the *NtrC* gene product, (NTRC) a positive control protein, together with RNA polymerase modified by σ^{54} activates transcription from the *nifLA* promoter. This cascade is involved in regulating the expression of a number of nitrogen-regulated enzyme systems in bacteria (See Dixon 1989).

NITROGEN FIXATION GENES IN AZOTOBACTER

The organisation of *nif* genes in Azotobacter is more complex than in *K. pneumoniae* (See Merrick 1988). *A. vinelandii* has a major *nif* gene cluster homologous to the Mo-*nif* gene cluster of *K. pneumoniae* but contains additional open reading frames of unknown function, and lacks *nifABQ* which map outside this region. This Mo-*nif* cluster of *A. vinelandii* has been sequenced (Jacobson *et al.* 1989) and a comparable region is present in the genome of *A. chroococcum* (Evans *et al.* 1988). In addition there is a cluster of structural genes encoding the V-nitrogenase (*Vnf* ≡ vanadium-dependent nitrogen fixation) (Robson *et al.* 1989; Joerger *et al.* 1990), and in *A. vinelandii* cluster of structural genes encoding the third nitrogenase (*anf* ≡ alternative nitrogen fixation) (Joerger *et al.* 1989a), Figure 2. The structural genes of V-nitrogenase and the third nitrogenase have also been sequenced and assigned to polypeptides of the purified proteins. As discussed below, a comparison of the derived amino acid sequences of these genes has provided a considerable insight as to the features likely to be important in binding the redox centres these proteins contain. In both these nitrogenase systems the larger component contains an additional subunit encoded by *VnfG* or *anfG* which has no known counterpart in the MoFe protein of Mo-nitrogenase. Upstream from the *anfHDGK* gene cluster an *anfA* gene has been identified.

Some *nif* genes appear to be required by both Mo-nitrogenase and Mo-independent nitrogenases, *nifB* and *nifM* (Kennedy *et al.* 1986) (both processing genes)

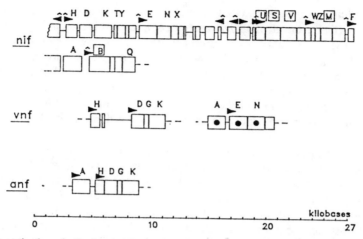

Fig. 2. Organisation of *nif vnf* and *anf* genes in *A. vinelandii*. Genes required for the function of all three nitrogenases are shown boxed (C. Kennedy unpublished, Bishop and Joerger 1990); < indicates -12-24 σ^{54} modified RNA polymerase recognition sequences (*rpo* N -dependent) and those promotors with upstream activating sequences (TGT N_{10} ACA) forming *nifA* binding sites. (For a review see Bishop and Joerger 1990).

are in this class, however, some of the other nitrogenase processing genes are reiterated e.g. a second *nifEN*-like sequence has recently been shown to be part of the *vnfAEN* cluster (Joerger *et al.* 1989b). The *nifA* gene and a regulatory gene *nfrX* with no known counterpart in *K. pneumoniae* is required for Mo-nitrogenase synthesis (Santero *et al.* 1988). *vnfA* is required for V-nitrogenase synthesis and the third nitrogenase requires *anfA* (Joerger *et al.* 1989b). The derived amino acid sequences of *nifA, vnfA,* and *anfA* are highly (37%-38%) conserved, indicating a role as positive activators for these proteins. How the activity of these proteins is modulated in Azotobacter is not known but all three systems are repressed in the presence of excess fixed-N.

Although as discussed above the regulation of *nif* expression by fixed-N has been the subject of intensive investigation, far less is know about the regulation of these systems by metals (See Pau 1989). Molybdenum has been shown to stimulate expression of the *nifHDKY* operon in *K. pneumoniae* and is required for transcription of this operon in *A. vinelandii*. Expression of the V-nitrogenase is repressed by Mo and the third nitrogenase is repressed by both Mo and V. V only prevents expression of the third nitrogenase if the structural genes of V-nitrogenase are present. (Luque and Pau 1990).

Biochemistry oF Nitrogen Fixation

Initial studies on the biochemistry of the involvement of vanadium in nitrogenase function in Azotobacter were undertaken before *nifHDK* deletion strains were available, and were inconclusive. In the early 1970s nitrogenase was isolated and partially-purified from *A. vinelandii* and *A. chroococcum* grown in the presence of vanadium (McKenna *et al.* 1970; Burns *et al.* 1971). These preparations showed an altered substrate specificity, but were of low specific activity and contained molybdenum. The difficulty of removing adventitious molybdenum from bacteriological medium to a level where organisms cannot accumulate it led to the suggestion that vanadium somehow 'spared' residual molybdenum in the growth medium, and thus the slow growth observed was Mo-dependent (Benneman *et al.* 1972).

The first unequivocal isolation of a V-nitrogenase was reported in 1986 from a strain of *A. chroococcum* in which the *nifHDK* genes were deleted (Robson *et al.* 1986) and subsequently from a comparable deletion strain of *A. vinelandii* grown in the presence of vanadium (Hales *et al.* 1986a), conditions which repress the expression of the third nitrogenase that this organism contains.

The V-nitrogenase of these organisms are sensitive to irreversible inactivation by dioxygen but they have been purified to homogeneity using conventional anaerobic techniques. They are comprised of a vanadium and iron-containing protein (VFe protein) (Hales *et al.* 1986a; Eady *et al.* 1987) and an iron-containing protein (Fe protein) (Hales *et al.* 1986b; Eady *et al.* 1988), both essential for nitrogenase activity. Although the preparations currently available are homogenous with regard to polypeptides, the variation in metal contents and specific activities indicates inactive or partially-active species lacking metals may be present. A similar situation is encountered with Mo-nitrogenase components.

To simplify the discussion of the components of nitrogenase of different types the protein components of the V-nitrogenase of *A. chroococcum* will be referred to as $Ac1^v$ for the VFe protein, and $Ac2^v$ for the Fe protein, to distinguish these proteins from the Mo-nitrogenase, $Ac1^{Mo}$ for the MoFe protein and $Ac2^{Mo}$ for the Fe protein. Similarly the components of V-nitrogenase of *A. vinelandii* are designated $Av1^v$ and $Av2^v$ to distinguish them from those of the Mo-nitrogenase $Av1^{Mo}$ and $Av2^{Mo}$ of this organism. The components of the Mo-nitrogenase of *K. pneumoniae* are designated as $Kp1^{Mo}$ and $Kp2^{Mo}$. This general nomenclature of Eady *et al.* (1972) is widely used in the nitrogenase field. The oxidation status of the proteins are indicated by the subscripts ox and red indicating oxidized or the dithionite-reduced species as necessary.

THE FE PROTEINS OF MO-NITROGENASE

The Fe proteins associated with Mo-nitrogenase are encoded by the *nifH* gene which has been cloned and sequenced from 20 or so diazotrophs ranging from Archebacteria through Eubacteria to Cyanobacteria. The derived amino acid sequence for Fe proteins from these studies show that the Fe proteins constitute a family of proteins with a structure which has been highly conserved in evolution (See Chen *et al.* 1986; Pretorius *et al.* 1987; Souillard *et al.* 1987). The percentage of identical amino acid sequence matches ranged from 45, comparing the archebacterium *Methanococcus thermolithotrophicus* with *Rhizobium sesbania*, to 97 when comparing two species of Rhizobia. The constraints placed on amino acid substitution in Fe proteins arise from a requirement to retain effective iron-sulphur cluster and nucleotide-binding sites, protein contact surfaces for stable subunit/subunit interactions, and the formation of transient protein complexes to allow electron transfer from a reduced donor, e.g. flavodoxin hydroquinone, and electron transfer to the MoFe protein. Features which are recognisable from the invariant regions common to all known *nifH* sequences are (1) the position of five invariant Cys residues which are candidates for Fe/S cluster ligation and have a proposed role in nucleotide binding, (2) the presence of a nucleotide binding motif at the NH_2-terminus. These characteristic sequences themselves are located in highly conserved surrounding regions, although the spacing of the conserved cysteinyl residues is not the same as is observed in simple Fe/S proteins.

Fe proteins of nine Mo-nitrogenase have been purified to apparent homogeneity (extensively reviewed in Burgess (1985); Eady (1986); Orme-Johnson (1985)). They are γ_2 dimers of approximate M_r of 62000 and contain approximately 4 Fe atoms and 4 S^{2-} atoms per dimer. The activity of the Fe proteins, even when isolated from aerobes, is very sensitive to inactivation by dioxygen (typical half lives are $\sim 45s$ in air).

The Fe and S^{2-} atoms present in Fe proteins have been assigned to a single [4Fe-4S] cluster on the basis of cluster extrusion data and comparison of various spectral parameters with FeLS proteins containing characterized [2Fe-2S] and [4Fe-4S] centres. The [4Fe-4S] centre traverses the 1 + and 2 + core oxidation states between the dithionite-reduced species and the dye-oxidized or catalytically-oxidized species present during enzyme turnover. The change in the visible absorption spectra associated with oxidation/reduction has been extensively used in kinetic studies.

It has been proposed that the [4Fe-4S] cluster bridges the two subunits. The nature of the ligands involved in binding the cluster in $Av2^{Mo}$ has been investigated by Fe K-edge X-ray absorption spectroscopy (Lindahl *et al.* 1987). The EXAFS data are consistent with each of the 4Fe atoms being ligated by 4 sulphur atoms at 2.31Å, and that most probably the cluster is bound by 4 cysteinyl sulphur residues. This assignment is supported by a site directed mutagenesis study of the effect of Cys→ Ser mutations in $Av2^{Mo}$ (Howard *et al.* 1989). The retention of Cys_{97} and Cys_{182} was essential for activity, consistent with an earlier chemical modification study which implicated these residues in both polypeptide chains in binding the [4Fe-4S] centre.

At low temperatures the dithionite-reduced Fe proteins exhibit complex EPR spectra since two interconvertible spin species are present. A rhombic EPR signal near g = 1.94 arising from an S = 1/2 spin system, and weaker signals at g $\sim$ 3.8 − 5.6 corresponding to an S = 3/2 ground spin state characterise this oxidation state. The sum of the spin intensities give approximately 1 spin per dimer, the ratio of the two is altered by the addition of urea or ethylene glycol. Although a ^1H-nmr study of the protons close to the FeS centre of $Cp2^{Mo}$ have been interpreted as the mixed species arising from an active site distortion consequent or freezing the EPR samples, their presence is unusual in naturally occurring FeS proteins (Meyer *et al.* 1988). The midpoint potential of the [4Fe-4S] cluster varies depending on the source of the Fe protein from -240mV to -390mV.

Fe proteins bind both MgADP and MgATP at two sites, MgADP inhibits the binding of MgATP. Quantitation of the binding constraint is difficult and for MgATP they range from 17 μM to 560 μM. The binding of nucleotides results in a protein conformational change and altered properties of the [4Fe-4S] cluster. Spectroscopically these changes consequent on the binding of MgATP result in the EPR signal at g = 1.94 becoming more axial, the ^{57}Fe quadrupole splitting of the Mossbauer spectrum decreases, and changes occur in the circular and magnetic circular dichroism spectra. The Fe protein becomes more susceptible to inactivation by dioxygen and the Fe of the FeS centre becomes more accessible to chelation by $\alpha\alpha'$ bipyridyl. The redox potential of the [4Fe-4S] cluster becomes $\sim$ 100mV more negative. Binding of MgADP results in a larger decrease in midpoint potential of -190mV in the case of Av2^{Mo}. MgADP is released slowly from the Ac2^{Mo} or Kp2^{Mo} nucleotide complexes and inhibits the rate of reduction of the oxidized protein by SO$_2{}^-$. Chemical modification experiments also implicate Cys$_{85}$ of Av2^{Mo} in binding nucleotides.

Fe Proteins of V-Nitrogenase

The structural genes of Ac2^V (Robson *et al.* 1986) and Av2^V (Raina *et al.* 1988; Joerger *et al.* 1990) have been cloned and sequenced. Comparison of the derived amino acid sequences of Ac2^V with Ac2^{Mo} (Robson *et al.* 1989) shows 89% homology with conservation of the invarient Cys residues and their spacing, and the nucleotide binding motif discussed above.

Both Av2^V and Ac2^V have been purified to homogeneity (Hales *et al.* 1986b; Eady *et al.* 1988). They are γ_2 dimers and contain approximately 4Fe atoms and 4 acid-labile S atoms per dimer (Table I). The dithionite-reduced species exhibit EPR spectra characteristic of the presence of both S = 1/2 and S = 3/2 spin states. In the presence of MgATP the rate of cheletion of Fe by $\alpha\alpha'$ bipyridyl is enhanced.

The E$_m$ values of Ac2^V.MgADP at -463mV and Ac2^{Mo}.MgADP at -450mV, and the rates of reduction of these oxidized species by dithionite in a 1 electron reduction are very similar (Bergstrom *et al.* 1988).

These data indicate structural homology with the electron transfer site associated with the [4Fe-4S] centre of Fe proteins of V and Mo-nitrogenase. This homology allows the formation of the functional heterologous nitrogenase between V and Mo-nitrogenase components, a property not always shown by components of Mo-nitrogense isolated from different organisms. It is clear that the Fe proteins of the V-nitrogenase of Azotobacter are typical members of the family of characterized Fe proteins associated with Mo-nitrogenase discussed above.

Table I. Comparison of the physicochemical properties of the Fe proteins of V-nitrogenase and Mo-nitrogenase.

	Ac2^v	Av2^v	Kp2^{Mo}
Native M_r	63000	63500	66800
Subunit structure	γ_2	γ_2	γ_2
Subunit M_r	32000	31000	34600
Metal and S^{2-}			
Content (g atom/Mol)			
Fe	3.7	3.4	4
S^{2-}	3.9	ND	3.85
EPR g-valve dithionite	2.035, 1.941,	2.05, 1.94,	2.05, 1.94
reduced	1.892	1.85	1.86
protein	4–5	4–5	4–5
Specific Activity			
nmol product/min/mg			
of protein			
NH$_3$ from N$_2$	337	ND	275
C$_2$H$_4$ from C$_2$H$_2$	341	1100–1400	980
H$_2$	1211	ND	1050
Oxygen			
Sensitivity, t$^{1/2}$	45 s	ND	45 s

References: Ac2^v (Eady et al. 1988); Av2^v, (Hales et al. 1986), Kp2^{Mo}, Eady et al. (1972) and Hagen et al. (1985).

The VFe Protein and MoFe Proteins

Since the discovery of V-nitrogenase, attention has focused on the structure of the VFe protein to establish the nature of the redox centres present, and the extent to which they are comparable with those of the MoFe proteins. The structure and spectroscopic properties of MoFe protein are reviewed in: Burgess (1984), Orme Johnson (1985), Stephens (1985) and Smith et al. (1987).

The consensus view of the structure of MoFe proteins is that they contain two types of unique redox centres within an $\alpha_2\beta_2$ tetramer. Preparations of MoFe protein with the highest activity contain 2 Mo atoms and 33 $\pm$ 3 Fe atoms and a marginally lower content of acid-labile sulphur atoms per tetramer. The Mo and approximately half the iron are present in a polynuclear cluster responsible for the unique EPR spectrum of the dithionite-reduced proteins. This spectrum, with g values near 4.3, 3.7 and 2.01 arises from an S = 3/2 spin system, is associated with an iron and molybdenum containing centre (FeMoco). FeMoco can be extracted from MoFe protein by treatment with N-methylformamide (NMF) and as described below there is good evidence that it is, as it forms a major part of, the active site at which substrates are reduced. The remainder of the Fe and an equivalent amount of S^{2-} can be extracted from the MoFe proteins as [4Fe-4S] clusters. In the intact MoFe protein these clusters (termed 'P' clusters) have unique spectroscopic properties. They have no counterpart in simpler Fe/S proteins and at -480 mV have the lowest redox potential of the centres present in $S_2O_4^{2-}$-reduced MoFe proteins. The role of the 'P' clusters is not clear. There is evidence for the formation of EPR detectable transients formed in turnover involving enzyme-bound substrate (C_2H_2) and inhibitor (CO) complexes but their function remains ill-defined at present.

The NMF extract containing FeMoco will combine spontaneously with puri-fied, but inactive MoFe protein, isolated from mutants unable to synthesise FeMoco, to form an active MoFe protein. This process has been most intensively studied with the $nifB^-$ Kpl protein of *K. pneumoniae*, which lacks FeMoco but contains the 'P' clusters. The most compelling evidence that FeMoco centres in the MoFe proteins forms part of the active site has provided studies on the transfer of unusual substrate reduction patterns associated with mutant MoFe proteins with FeMoco to the reactivated $nifB^-$ Kpl. Specifically, mutations in the $nifV$ gene result in the synthesis of MoFe protein for which C_2H_2 is an effective substrate but N_2 is not, and unlike the wild-type enzyme CO inhibits hydrogen evolution. When FeMoco is extracted from purified $nifV^-$ Kpl in NMF and recombined with $nifB^-$ Kpl which lacks FeMoco, the resulting protein has the substrate specificity and altered inhibition pattern normally associated with the $nifV$ mutation. Recent work suggests that this change is associated with the replacement of homocitrate as a component of FeMoco by citrate (See Hoover *et al.* 1988 for review). As described below, this type of experiment, i.e. the extraction of the VFe proteins with NMF and the restoration of activity to $nifB^-$ Kpl to form a hybrid enzyme with the characteristic substrate reduction pattern of VFe protein, provides strong evidence for the presence of an Fe and V containing cofactor centre, analogous to FeMoco of the MoFe proteins in the VFe proteins.

The VFe proteins of *A. chroococcum* and *A. vinelandii* have been purified and characterised. Their overall physicochemical properties are similar to those of MoFe proteins, having homologous α and β subunits, except that V replaces Mo. (Table II). The lower Fe content of preparations of VFe protein currently available when compared with that of MoFe proteins is probably not significant since experience with MoFe protein would suggest that they will increase with man-years of effort!

The structural genes encoding the polypeptides of the V-nitrogenase of *A. chroococcum* have been cloned and sequenced (Robson *et al.* 1989). The NH_2-ter-minal amino acid sequence of the subunit polypeptides of the VFe protein separated by SDS electrophoresis have been determined and comparison of this sequence with that derived from the nucleotide sequence has allowed the assignment of genes to polypeptides. This approach resulted in the identification of an additional type of small subunit in AcI^V (the δ subunit) not found in MoFe proteins and encoded by the $vnfG$ gene (see Figure 3). The δ subunit has not been described for the VFe protein of *A. vinelandii* but the presence of an open reading frame with homology to $vnfG$ of *A. chroococcum* (Joerger *et al.* 1990), makes it highly likely that both proteins contain the additional subunit and have an $\alpha_2\beta_2\delta_2$ subunit structure.

The amino acid sequences of the polypeptide subunits of MoFe proteins

Table II. Comparison of the physicochemical properties of VFe proteins and MoFe proteins.

	$Ac1^V$	$Av1^V$	$Kp1^{Mo}$
Native M_r	210000	200000	220000
Subunit structure	$\alpha_2\beta_2\gamma_2$	$\alpha_2\beta_2(\gamma_2)$	$\alpha_2\beta_2$
Subunit M_r	2×50000	2×52000	2×50000
	2×55000	2×55000	2×60000
	2×14000		
Metal and S^{2-} content (g atom/mol)			
V	$2 \pm .3$	$0.7 \pm .3$	ND
Mo	0.06	0.05	2
Fe	19 ± 2	9 ± 2	32 ± 3
S^{2-}	20 ± 2	ND	ND
EPR g-valve dithionite	5.6 4.35 3.77	5.31 4.34	4.3 3.7
reduced protein	1.93	2.04 1.93	2.01
Specific Activity nmol protein min/mg of protein			
NH_3 from N_2	350	660	990
C_2H_4 from C_2H_2	608	220	1693
H_2	1348	1400	2100
Oxygen sensitivity, $t_{1/2}$	45 s	ND	10 min

References: $Ac1^V$ Eady *et al.* (1987), $Av1^V$ Hales *et al.* (1986), $Kp1^{Mo}$ Eady *et al.* (1972)

encoded by *nifD* and *nifK*, are highly conserved among diazotrophs of different taxonomic groups. In all known sequences there are five invariant Cys residues in the α subunit and three in the β subunit (Figure 4). Although these residues do not show the characteristic Cys X X Cys spacing of simple Fe S proteins they have been proposed to have a role in ligating the FeMoco and 'P' cluster redox centres present in these proteins. These invariant Cys clusters are also conserved in the

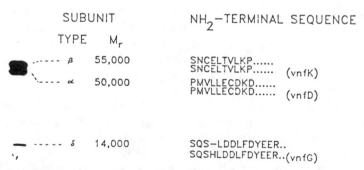

	SUBUNIT TYPE	M_r	NH_2-TERMINAL SEQUENCE	
	β	55,000	SNCELTVLKP...... / SNCELTVLKP......	(vnfK)
	α	50,000	PMVLLECDKD...... / PMVLLECDKD......	(vnfD)
	δ	14,000	SQS—LDDLFDYEER.. / SQSHLDDLFDYEER..	(vnfG)

Fig. 3. Subunit structure of the VFe protein. Polypeptides of $Ac1^V$ were separated by SDS poly-acrylanide gel electrophoresis and show three bands. The NH_2 terminal amino acid sequences of each polypeptide are compared with the predicted sequences of the *vnf* structural genes, and confirm a gene *vnfG* encoding the α subunit of $Ac1^V$ (Robson *et al.* 1989).

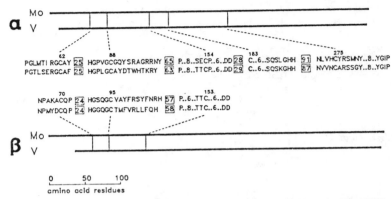

Fig. 4. Schematic representation of the α and polypeptide subunits of MoFe protein and VFe protein. The vertical lines indicate the position of invarient Cys residues of Ac1ᵛ and Av1ᴹᵒ. The flanking amino acid sequences are shown as is His 185 of the β subunit which has the same special relationship as Cys 183 to the other conserved Cys residues of the α subunit. The numbering is from reference to AV1ᴹᵒ and the boxed numbers are inter cysteinyl or inter cysteinyl-histidine spacing. The subunits of nitrogenase-3 of *A. vinelandii* show a similar sequence relationship (see Pau 1989).

homologous *vnfD* and *vnfK* genes. However, there are insufficient conserved Cys residues to allow conventional ligation patterns to the 'P' clusters even ignoring Cys ligation to FeMoco in MoFe proteins. Since site directed mutational studies have implicated Cys residues in binding the latter to MoFe proteins, it is highly likely that noncystinyl ligation is involved in redox cluster binding to the MoFe proteins, and since like regions may be supposed to have similar functions, to the VFe proteins.

VFE PROTEIN-SPECTROSCOPIC STUDIES

As described below, EPR, MCD and V K-edge X-ray absorption spectroscopic studies are consistent with the VFe proteins containing the same types of redox centres as are found in MoFe proteins. In addition, the $S_2O_4^{2-}$-reduced proteins exhibit two EPR detectable species which have no counterpart in MoFe proteins.

ELECTRON PARAMAGNETIC RESONANCE SPECTROSCOPY

In the case of $S_2O_4^{2-}$-reduced MoFe proteins, EPR spectra are characterised by a spin S = 3/2 signal, arising from the FeMoco centres, with g values near 4.3, 3.7 and 2.01, and integrating to 0.94 spin/Mo atom. The spectra of VFe proteins are more complex since several paramagnetic centres are detectable.

The VFe protein of *A. vinelandii* shows a broad, poorly resolved signal centred at g = 5.5 (Morningstar and Hales, 1987). This signal has been assigned to the low field inflexion of the transition from the ground state to the first excited state of the two Kramer's doublets of a spin S = 3/2 system, and integrates to 0.89

spin/V atom. The VFe protein from *A. chroococcum* has EPR signals assignable to a S = 3/2 ground state spin system with g values at 5.6, 4.35 and 3.77 (Eady *et al.* 1987). However, this signal (Figure 5) has proved difficult to simulate and may arise from a mixture of species reflecting different environments of the cofactor centre which these features of the EPR spectrum suggest is present (Lowe, personal comm.).

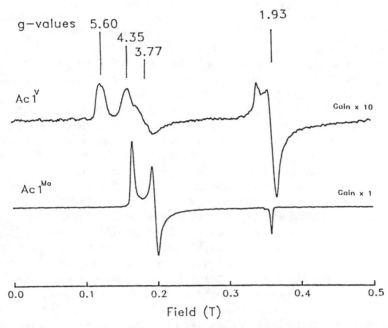

Fig. 5. EPR spectra of Ac1^V and Ac1^{Mo}. The spectra were run at 10K with a 0.1mT field modulation at 100 KHz and a microwave power of 20.6mW at 9.85GHz. The dithionite-reduced proteins were Ac1^V (17mg/ml) and Ac1^M (12mg/ml). Note the 10-fold difference in gain between the two spectra.

The EPR spectra of VFe proteins from both organisms shows an axial signal with g values 2.04 and 1.93 assigned to an S = 1/2 ground state spin system. This signal which is not present in spectra of MoFe proteins probably arises from a reduced Fe/S centre and integrates to 0.2 – 0.3 spin/mol. On dye-oxidation this signal is bleached before the S = 3/2 species and presumably has a lower redox potential.

A third EPR species has recently been reported to be present in spectra of the VFe protein from *A. vinelandii* (Hales *et al.* 1989). This signal, which is difficult to detect in the absorption mode is detectable in the dispersion mode. At low temperature (2K) and high power to saturate the resonances to bring them into passage, a broad structureless absorption envelope predominates the spectrum. The signal which extends from 2.8KG to ~ 5.5KG with a peak at 3.5KG has a non-classical line shape and is characteristic of a paramagnetic site coupled to a metal centre. This signal is also not detectable in the spectra of MoFe proteins.

MAGNETIC CIRCULAR DICHROISM SPECTROSCOPY

Low temperature MCD spectroscopy has proved to be a powerful technique when applied to the MoFe proteins since spectral features arising from paramagnetic centres are temperature dependent and those of diamagnetic species are not. At low temperature features arising from paramagnetic centres predominate the spectra. For $S_2O_4^{2-}$-reduced MoFe protein where Mossbauer studies indicate that the FeMoco centres are the only paramagnetic species, the MCD spectrum is sufficiently unique as to provide an optical fingerprint for this cluster in its $S_2O_4^{2-}$-reduced state. The magnetic dependence of the spectrum is consistent with the EPR assignment of an S = 3/2 spin system.

Oxidation of MoFe proteins with an excess of the redox dye thionine, results in oxidation of both the 'P' clusters and the FeMoco centres without loss of activity. In the EPR silent form of the oxidized protein, Mossbauer spectroscopy indicates that the 'P' clusters are the only paramagnetic centres and the MCD spectra that they are spin systems with S = 5/2 or S = 7/2 spin states. The magnetization curves of the MCD transitions of the 'P' clusters are uniquely steep in comparison with conventional Fe/S centres and these unusual properties may arise from noncysteinyl ligation to the clusters as described above.

MCD studies on VFe proteins are restricted to Av1^v (Morningstar et al. 1987). Although the EPR spectra of the $S_2O_4^{2-}$-reduced protein exhibits three distinct paramagnetic species, the MCD spectra have been amenable to analysis. Compared with the MoFe proteins there are marked differences in the intensity and frequency of the temperature-dependent MCD transitions. However magnetization curves at 800nm and 520nm are quantitatively similar and the pronounced nesting indicate that the predominant MCD transitions of Av1^v arise from the S = 3/2 spin species with only minor contributions from the other paramagnetic centres. The differences between these spectra and those of the MoFe proteins were attributed to different zero field splitting parameters of the centres in these proteins. However this data by comparison with FeMoco provided good presumptive evidence for the presence of a vanadium containing cluster in the VFe proteins.

Thionine oxidation of Av1^v bleaches the EPR signals associated with the paramagnetic species in $S_2O_4^{2-}$-reduced protein. The oxidized Av1^v has low temperature MCD spectra similar but not identical with those of oxidized MoFe proteins. Although the spectra are 2-3 fold less intense in the case of the VFe protein, their form and the steepness of the magnetization curves together with the absence of any EPR signals has been interpreted as their originating from Fe/S clusters with similar electronic and magnetic properties as the "P" clusters of the MoFe proteins.

X-RAY ABSORPTION SPECTROSCOPY

Crystals of the MoFe protein suitable for X-ray analysis have been available for some time, but despite considerable effort no detailed information as to the structure, and in particular the environment of Mo, are available. In contrast, metal X-ray absorption spectroscopy has provided a detailed picture of the environment of Mo and V in nitrogenase systems.

The absorption edge and X-ray absorption near-edge structure (XANES) region of an X-ray absorption spectrum contains information about coordination geometry and probable oxidation level, and the extended X-ray absorption fine structure (EXAFS) the distance type and number of nearest neighbour atoms to the primary absorbing atom.

The VFe proteins of both *A. chroococcum* (Arber *et al.* 1987, 1989) and *A. vinelandii* (George *et al.* 1988) have been studied by vanadium K-edge X-ray absorption spectroscopy and the results are in good agreement. The spectra of $S_2O_4^{2-}$-reduced proteins show an edge position and pre-edge feature indicating an oxidation state between V^{II} and V^{IV}. The very weak pre-edge feature (7% of the edge height) is consistent with distorted octehedral coordination of the V atom. These data closely resemble those for the octahedral VFe_3S_4 complex in $[NMe_3][VFe_3S_4(DMF)_3]^-$ (Figure 6), and are clearly different from the complex $[VS_4(FeCl_2)_2][Et_4N]_3$ where vanadium is tetrahedrally coordinated to sulphur.

The vanadium edge and XANES region of $Ac1^V$ do not change significantly

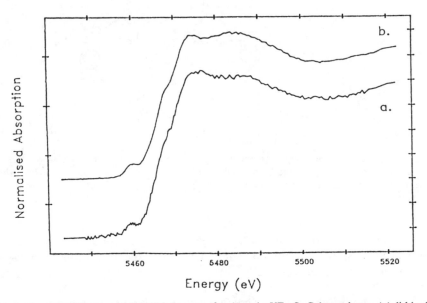

Fig. 6. Vanadium K-edge and XANES Spectra of $Ac1^V$ and a VFe_3S_4 Cubane cluster. (a) dithionite-reduced $Ac1^V$ (b) $[VFe_3S_4Cl_3(DMF)_3][NMe_4]^-$.

when the protein is oxidized with thionine a 'super reduced' as occurs in enzyme turnover. The very small differences between these spectra can be explained by slight variation in the vanadium environment with angular rather than radial changes occurring (Arber *et al.* 1989).

Fourier transform analysis of the EXAFS region of the spectra of both $S_2O_4^{2-}$-reduced VFe proteins are consistent with the vanadium being part of a V-Fe-S cluster. The analysis is consistent with 3 ± 1 sulphur atoms at 2.31(3)Å, 3 ± 1 iron atoms at 2.75(3)Å and there is evidence for *ca*, three light atoms (oxygen, nitrogen) at 2.15(3)Å. A study of dye-oxidised AcI^V showed no significant changes in the EXAFS spectrum. These dimensions and types of atom are very similar to those derived by X-ray crystallographic structure analysis of the cluster compound $[NMe_3][VFe_3S_4(DMF)_3]^-$ (Kovacs and Holm 1986) (Figure 7). Comparison of

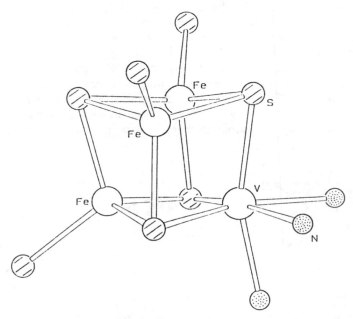

Fig. 7. The structure of the complex $[MeN_4][VFe_3S_4Cl_3(DMF)_3]$. This complex was synthesized and the X-ray structure determined by Kovacks and Holm (1986). The V atom in this cluster has a chemical environment very similar to that of V in VFe protein. (Arber *et al.* 1987; George *et al.* 1988) (See Table III).

these data with those of a study of Mo K-edge X-ray absorption spectroscopy of MoFe proteins shows the environment of V in VFe proteins to be very similar to those of Mo in MoFe proteins and indicate that V is part of a polynuclear cluster analogous to FeMoco.

As described below, a cofactor containing VFe and S has been extracted from AcI^V. This cofactor has been studied by iron K-edge X-ray absorption spectroscopy and EXAFS data provide evidence for the presence of FeS_2M (M = V or Fe) rhombs (Garner *et al.* 1989). Two Fe – Fe separations are present, one set at 2.61Å and a longer one at 3.65Å. Similar data were obtained for FeMoco (Arber *et al.* 1988) and indicate that the metal-sulphur framework of both cofactors is sufficiently ordered as to maintain coherent Fe – Fe backscattering contributions. These spectroscopic data emphasize the similarity in the chemical environment of Mo and V in these nitrogenase systems (See Table III).

Table III. Comparison of the environment of V in VFe proteins with Mo in MoFe proteins determined from X-ray absorption spectroscopy.

	Vanadium environment			
	Type of atom	No.	R (Å)	
$Ac1^V$	V-O (or C or N)	3 ± 1	2.15	
	V-S	3 ± 1	2.31	
	V-Fe	3 ± 1	2.75	
$Av1^V$	V-O (or C or N)	2–3	2.15	
	V-S	3–4	2.33	
	V-Fe	3 ± 1	2.76	
			EXAFS	X-ray crystallographic analysis
$[Me_4][VFe_3S_4Cl_3(DMF)_3]$	V-O	3	2.12	2.13
	V-S	3	2.35	2.33
	V-Fe	3	2.75	2.77
	Molybdenum environment			
$KP1^{Mo}$	Mo-O(or C or N)	12.12		
	Mo-S	4.5 ± 1	2.37	
	Mo-Fe	3.5 ± 1	2.67	
$Av1^{Mo}$	Mo-O(or C or N)	1.9	2.12	
	Mo-S	4.5	2.37	
	Mo-Fe	3.5	2.67	

References: $Ac1^V$ (Arber *et al.* 1987, 1988), $Av1^V$ (George *et al.* 1988), $[VFe_3S_4Cl_3(DMF)_3]^-[Me_4]$, (Kovacks and Holm 1986; Arber *et al.* 1987), $KP1^{Mo}$ (Eidsness *et al.* 1986), $Av1^{Mo}$ (Newton *et al.* 1984)

THE VANADIUM AND IRON CONTAINING COFACTOR

The spectral data discussed above clearly indicate the chemical environment of V in VFe proteins is similar to that of Mo in MoFe proteins. That the V is present in a cofactor centre in the VFe protein was shown by extraction of $Ac1^V$ with

N-methylformamide (NMF) under conditions similar to those used to extract FeMoco from MoFe proteins (Smith *et al.* 1988) (see Figure 8). The NMF extract was brown and contained V, Fe and acid-labile sulphur in the approximate ratio 1 : 6 : 5. EPR spectra exhibited weak signals characteristic of an S = 3/2 metal-containing centre.

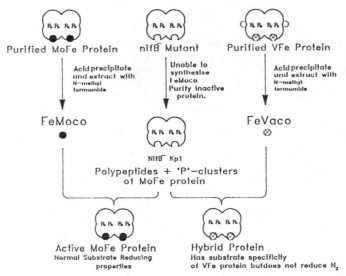

Fig. 8. Extraction and assay of iron molybdenum cofactor from MoFe proteins and iron vanadium cofactor from VFe protein.

When combined with purified nifB⁻ Kpl the NMF extract activated the MoFe protein polypeptides and conferred the substrate specificity of Acl^V, i.e. the formation of C_2H_6 as a product of C_2H_2 reduction. The transfer of this pattern strongly suggests that the polynuclear Fe, V and S containing centre (FeVaco) had been extracted from the VFe polypeptides without major disruption of the cluster. This result is consistent with the Fe K-edge EXAFS spectra discussed above.

The hybrid protein, although active in C_2H_2 reduction and H^+ reduction assays, could not reduce N_2 to NH_3 or N_2H_4. This observation suggests that the reduction of N_2 requires specific interactions of polypeptide ligands with the cofactor centre, and that these are missing in the FeVaco/nifB⁻Kpl protein (Smith *et al.* 1988).

SUBSTRATE REDUCTION

Like Mo-nitrogenase, V-nitrogenase requires MgATP and a low potential reductant ($S_2O_4{}^{2-}$) to reduce N_2 or C_2H_2. In the case of Mo-nitrogenase, extensive presteady-state spectroscopic and kinetic studies are consistent with the Fe protein acting as an ATP-dependent electron donor to the MoFe protein which contains the substrate binding site. Electron transfer between the two proteins is tightly coupled to the hydrolysis of MgATP, and the product MgADP is a potent inhibitor of electron transfer and substrate reduction. This area of research has been reviewed extensively (Lowe *et al.* 1985; Burgess 1985; Smith *et al.* 1987) and these sources contain the primary references for Mo-nitrogenases.

The functional similarity between Mo and V-nitrogenase component proteins is shown by the ability of $Ac1^{Mo}$ and $Ac2^v$ to form a fully-active nitrogenase, and $Ac1^v$ with $Ac2^{Mo}$ a nitrogenase with 70% the activity of the homologous system. The substrate reduction patterns for V-nitrogenase components are also similar to those of Mo-nitrogenase (Eady *et al.* 1987). Titration curves for the complementation of $Ac1^v$ activity with increasing concentrations of $Ac2^v$ for H^+, C_2H_2 and N_2 reduction show a rapid increase in activity up to an approximately 20-fold molar excess of $Ac2^v$, and thereafter only a slow increase. For the titration of $Ac1^v$ with $Ac2^{Mo}$ activity is 30% lower over the entire range indicating that the inefficient coupling is not overcome at high $Ac2^{Mo}$ concentrations. Titration curves for the complementation of $Ac2^v$ activity by $Ac1^v$ and $Ac1^{Mo}$ are the same and show an optimum at a 1:1 molar ratio with $Ac2^v$. As discussed below a notable difference is the relatively high rate of H_2 evolution in the presence of N_2 or C_2H_2 as reducible substrates by V-nitrogenase assayed at 30 °C.

REDUCTION OF N_2 AND H^+

The rate of electron flux through Mo-nitrogenase is essentially independent of the substrate being reduced, but reducing equivalents are partitioned depending on the rate of turnover and the reducible substrates available to the enzyme. During the turnover under an inert argon atmosphere protons are reduced to H_2. Under the same conditions, except under an atmosphere of N_2, H_2 evolution occurs to a lesser extent and the balance of the electron flux appears in NH_3.

Under conditions optimum for N_2 reduction, 25% of the electron flux is directed to H_2 formation and the stoicheiometry for N_2 reduction is:

$$N_2 + 8e + 8H^+ + 16\,MgATP \rightarrow 2NH_3 + H_2 + 16\,MgADP + 16\,Pi \quad (1)$$

H_2 evolution to this extent is still observed under 50 atm of N_2, and is thought to be of mechanistic significance in enzymic N_2 reduction, as dicussed below, rather than a side reaction resulting in a leakage of reducing equivalents.

A feature of V-nitrogenase of *A. chroococcum* is the decreased effectiveness with which N_2 competes as a reducible substrate with protons, as compared with Mo-nitrogenase. Although the apparent K_m for N_2 is similar (14KPa) for both nitrogenase systems, under an atmosphere of N_2 50% of the electron flux through V-nitrogenase is utilised for H^+ reduction. For Mo-nitrogenase this situation is only observed when N_2 is not staturating, or when the rate of electron transfer from the Fe protein to the MoFe protein is suboptional, as occurs at limiting Fe protein, MgATP or $S_2O_4^{2-}$ concentrations. Under these conditions additional H_2 is formed by the relaxation of the two-electron reduced species of MoFe protein formed in the initial stages of the N_2 reduction cycle, resulting in oxidation by H_2 evolution before N_2 can bind (see below and Scheme 1).

The higher electron allocation to H^+ reduction by V-nitrogenase in the presence of N_2 does not appear to be due to restricted electron transfer between the two proteins. A presteady state Kinetic study of the initial electron transfer reaction between $Ac2^v$ and $Ac1^v$ showed many similarities with this reaction of Mo-nitrogenase (Thorneley et al. 1989). Although the rate of electron transfer ($K_{+2} = 46 \pm 2 \ s^{-1}$) was found to be four times slower than for Mo-nitrogenase, it is still ten fold faster than the rate-limiting step in the catalytic cyle for substrate reduction. The overall similarity suggests that dissociation of oxidized Fe protein with MgADP bound may be rate-limiting for both Mo and V-nitrogenase and that differences in substrate specificity are a consequence of different reactivities at the FeMoco and FeVaco centres these proteins contain.

The continued reduction of protons by V-nitrogenase in the presence of N_2 makes this system less efficient in the utilization of reducing equivalents than Mo-nitrogenase, and the question arises as to why this apparently inefficient system should have persisted in Nature. However, as the assay temperature is lowered, N_2 continues to be reduced by V-nitrogenase whereas Mo-nitrogenase diverts more electron flux into H^+ reduction at the expense of N_2 reduction. At 5 °C the low activity of Mo-nitrogenase towards N_2 cannot be overcome by increasing the electron flux, and in the temperature range 30 °C to 5 °C the specific activity of $Ac1^{Mo}$ for N_2 reduction decreases 10-fold more than for $Ac1^V$ (Miller and Eady 1989a).

REDUCTION OF C_2H_2

C_2H_2 is a relatively poor substrate for V-nitrogenase due to the continued reduction of H^+ in the presence of C_2H_2 (Hales et al. 1986a; Eady et al. 1987). Under 0.1 atm of C_2H_2 the rates of formation of the 2 electron reduction products H_2 and C_2H_4 indicate that 60% of the electron flux is used for H_2 formation. Under similar conditions Mo-nitrogenase of K. pneumoniae, C_2H_2 reduced more effectively and only 25% of the electron flux goes to form H_2. The failure of C_2H_2 to compete with H^+ for V-nitrogenase is associated with the VFe protein, and not due to a higher apparent K_m for C_2H_2, since at 5.9 KPa it is close to the range observed for Mo-nitrogenase (0.6 to 2 KPa).

The reduction of C_2H_2 is highly stereospecific, giving essentially [cis $^2H_2$] ethylene when the reaction is carried out in ^2H_2O (Dilworth et al. 1988). A similar stereospecificity is shown by Mo-nitrogenase and has been widely interpreted as indicating side-on bonding of C_2H_2 at a metal site.

In addition to C_2H_4, V-nitrogenase forms C_2H_6 as a minor product of C_2H_2 reduction. This reaction product is not shown by Mo-nitrogenase, and assays using mixed Mo and V-nitrogenase components have shown C_2H_6 formation to be a property of the VFe protein (Dilworth et al. 1987, 1988). The apparent K_m for C_2H_2 reduction leading to both C_2H_4 and C_2H_6 is the same (5.9KPa) and the electron-flux-to-C_2H_4 ratio (C_2H_6/C_2H_4 ratio) remains at 2.35% over a range of

C_2H_2 partial pressures from 2 to 30 KPa. The C_2H_6/C_2H_2 ratio increases with increasing electron flux resulting from high $Ac2^v : Ac1^v$ ratios to 5.5%.

The time course for C_2H_6 formation shows a lag period of several minutes before a linear rate of product formation is observed. In contrast, the concomitant rate of C_2H_4 formation and H_2 evolution show no detectable lag. The slow accumulation of the intermediate leading to C_2H_6 formation cannot be explained by the Lowe Thorneley model for nitrogenase function. It has been shown not to be due to the accumulation of free C_2H_4 which is then further reduced, and different routes of reduction for C_2H_2 leading to C_2H_4 and C_2H_6 as products, have been proposed.

The distinctive property of V-nitrogenase in forming C_2H_6 as a reduction product of C_2H_2 has provided strong supportive evidence that the VFe protein contains a cofactor centre analogous to FeMoco (but containing vanadium) at the substrate-binding site. This reaction is also shown by cultures of azotobacter when grown under Mo-limitation and it has been proposed as a potential test for the functioning of Mo-independent nitrogenase *in vivo* (Dilworth *et al.* 1987). The ability of V-supplemented, but not Mo-supplemented cultures of *Clostridium pasterianum* (Dilworth *et al.* 1987) and the cyanobacteria *Anabaena variabilis* (Kentemitch *et al.* 1988) to form C_2H_6 from C_2H_2 make it likely that V-nitrogenase is not restricted to azotobacters.

REDUCTION OF OTHER SUBSTRATES

Ethylene has been shown to be reduced by Mo and V-nitrogenase to C_2H_6 (Ashby *et al.* 1987; Dilworth *et al.* 1988). Reduction by V-nitrogenase is slow, with an apparent K_m of 172 KPa compared with 130 KPa for the Mo-nitrogenase of *K. pneumoniae*. In contrast to C_2H_6 formation from C_2H_2 (discussed above), the time course for the reduction of C_2H_4 to C_2H_6 shows no detectable lag.

Cyanamide ($N \equiv CNH_2$) has recently been shown to be a substrate for both Mo and V-nitrogenase (Miller and Eady, 1989b). In both cases 6 and 8 electron reduction patterns are observed:

$$N \equiv CNH_2 + 6e + 6H^+ \rightarrow CH_3NH_2 + NH_3 \qquad (2)$$
$$N \equiv CNH_2 + 8e + 8H^+ \rightarrow CH_4 + 2NH_3 \qquad (3)$$

Which of these pathways predominates depends on the substrate concentration and on the ratio of nitrogenase components. Cyanamide is not such a good substrate for V-nitrogenase, the apparent Km at 2.6mM is 4-fold higher and the Vmax 8-fold lower than for Mo-nitrogenase. X-ray structural analysis of the molybdenum cyanamide complex *trans*[$Mo(NCN)_2(Ph_2PCH_2CH_2PPh_2)_2$] indicates that, in this complex binding occurs as a cyano-imide ligand (Hughes *et al.* 1988).

Propyne ($CH_3C \equiv CH$) is not reduced by V-nitrogenase, in contrast with Mo-nitrogenase where propyne (apparent $K_m = 80$ KPa) is reduced to propene. The inability to detect reduction by V-nitrogenase may be a reflection of the generally weaker affinity of this system for hydrocarbon substrates. If the affinity changed in the same ratio as for C_2H_2 an apparent K_m for propyne of the order of 300 KPa would be expected for V-nitrogenase (Dilworth *et al.* 1988).

INHIBITORS OF SUBSTRATE REDUCTION

With Mo-nitrogenase H_2 is a specific competitive inhibitor of N_2, but does not affect the reduction of other substrates, whereas CO inhibits the reduction of all substrates except H^+.

There have been no systematic studies on these inhibitors of V-nitrogenase but the overall pattern is similar to Mo-nitrogenase (Dilworth *et al.* 1988). H_2 (85 KPa) completely inhibited N_2 reduction by V-nitrogenase but had no effect on the reduction of C_2H_2 to either C_2H_4 or C_2H_6. CO (5 KPa) did not inhibit H_2 evolution by V-nitrogenase, but inhibited C_2H_2 reduction to C_2H_4 or C_2H_6 by 80%. However, the CO concentration required to produce 50% inhibition was $\sim$ 50-fold higher for V-nitrogenase compared with Mo-nitrogenase, consistent with metal carbonyl chemistry where weaker binding of CO to V compared with Mo is expected.

THE CATALYTIC CYCLE OF NITROGEN REDUCTION

The efficiency of substrate reduction by nitrogenase can be assessed from the determination of the ratio of moles of MgATP hydrolysed/mole H_2 produced (the ATP/2e ratio). For Mo-nitrogenase the ATP/2e ratio can vary considerably depending on the electron flux through the enzyme, but under optimum conditions values for this ratio are in the range 4.5-5.5. For V-nitrogenase a value of 5.5 was found for the ATP/2e ratio indicating that the efficiency of ATP utilization for substrate reduction by both systems is similar.

The most comprehensive model for Mo-nitrogenase function is that of Thorneley and Lowe developed from studies of the Mo-nitrogenase of *K. pneumoniae* (comprehensively reviewed in Thorneley and Lowe, 1985). Pre-steady state studies on the rate of the initial electron transfer between $Kp2^{Mo}$ and $Kp1^{Mo}$ using stopped-flow absorption spectroscopy, and rapid chemical-quench techniques allowed quantification of the rates of formation of an enzyme-bound intermediate present during N_2 reduction, with the rates of H_2 evolution and NH_3 formation. The model, which was developed from computer simulation of the complex kinetic data, can simulate all available data on N_2 reduction by Mo-nitrogenase and is consistent with the FeMoco centres acting independently.

In this model, each transfer of an electron from the Fe protein to the MoFe protein is considered as a one-electron cycle in which a reduced Fe protein-2MgATP species complexes with, and reduces the MoFe protein with concommitant hydrolysis of MgATP. This electron transfer reaction is rapid compared with the rate-limiting reaction. The protein-protein complex then dissociates in the rate-limiting reaction of the cycle, which is then completed by the release of MgADP from the Fe protein which is then re-reduced. During the reduction of N_2 eight such cycles occur sequentially. The initial electron transfer and protonation form a hydride species of the enzyme. After the transfer of three electrons to the MoFe protein, N_2 can bind to displace H_2 and accounting for the stoichiometry of eqn (1). Further electron transfer and protonation results in the formation of an enzyme-bound dinitrogen hydride intermediate which gives rise to N_2H_4 on chemical quenching of the enzyme, or to form NH_3 following further 4 electron transfers (see scheme 1). Comparison with the chemistry of NH_3

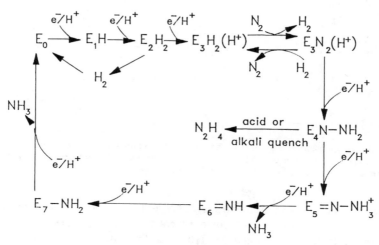

SCHEME 1

The catalytic cycle for N_2 reduction by Mo-nitrogenase of *K pneumoniae* E_o to E_7 represent intermediate forms of Kp1 after sequential reductions by one-electron. The arrows between each state represent the reactions of complex formation between Kp2 and Kp1, electron transfer between the two proteins and protonation followed by protein dissociation. N_2 binds to species E_3 to displace H_2, accounting for the stoichiometry of eqn (1). H_2 can displace N_2 from this species accounting for the competitive inhibition of N_2 reduction by H_2 (See Thorneley and Lowe 1985).

formation from N_2 at transition metal centres favour a mechanism in which the β N atom is progessively protonated as multiple bond character builds between the β N atom at a metal atom (putitively Mo) at the substrate-binding site. During N_2 reduction no *free* N_2H_4 is detectable and reduction proceeds through to NH_3 on the enzyme.

For V-nitrogenase the pre-steady state data are restricted to reactions involving the initial electron transfer reactions between $Ac2^V$ and $Ac1^V$ (Thorneley *et al.* 1989). The rate of electron transfer was measured by stopped-flow absorbance spectroscopy utilizing the increase in absorbance of the Fe protein as it becomes oxidized in enzyme turnover. The data were analysed in terms of scheme 2 analogous to that used for Mo-nitrogenase. The dependance of the rate of electron transfer on MgATP concentration and the competitive inhibition by MgADP gave K_D of 230 $\pm$ 10 μM for binding of MgATP and a K_i for MgADP of 30 $\pm$ 5 μM. A comparison of the dependance of the first order rate constant for electron

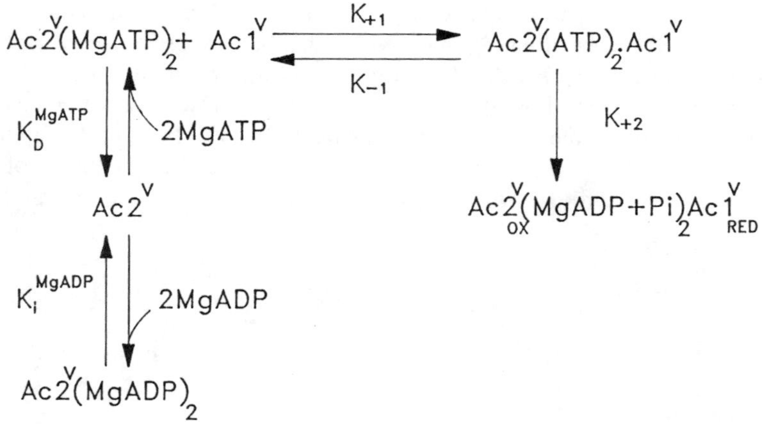

R.R. EADY

$$\text{Ac}\,2^V(\text{MgATP})_2 + \text{Ac}\,1^V \underset{K_{-1}}{\overset{K_{+1}}{\rightleftharpoons}} \text{Ac}\,2^V(\text{ATP})_2 \text{Ac}\,1^V$$

Scheme diagram:

Ac2^V(MgATP)$_2$ + Ac1^V ⇌ (K$_{+1}$ / K$_{-1}$) Ac2^V(ATP)$_2$Ac1^V

K$_D^{MgATP}$ ↕ ↘ 2MgATP — K$_{+2}$ ↓

Ac2^V — Ac2^V_{OX}(MgADP+Pi)$_2$Ac1^V_{RED}

K$_i^{MgADP}$ ↕ 2MgADP

Ac2^V(MgADP)$_2$

SCHEME 2

Mg ATP-dependent electron transfer from Ac2^V to Ac1^V and inhibition of this reaction by MgADP. The pre-steady state electron transfer between the components of V-nitrogenase of *A chrococcum* have been analysed in terms of this scheme (Thorneley *et al.* 1989).

transfer (K_{obs}) on protein concentration of V and Mo-nitrogenase indicates that the components of V-nitrogenase form a weaker electron transfer complex than do components of Mo-nitrogenase. However the rates of association of the components and the rate of MgATP-dependent electron transfer are similar in both systems. These reactions, which constitute the first two steps of the catalytic cycle of nitrogenase function described above suggest a similar mechanism for N_2 reduction by both types of nitrogenase.

CHEMICAL STUDIES ON THE REDUCTION OF N_2

Dinitrogen binds to most transition metals to form stable isolable compounds, vanadium was an exception until 1989 when the bridging dinitrogen complex [{V(2-Me$_2$NCH$_2$C$_6$H$_4$)$_2$(Py)}$_2$(μ-N$_2$)]. 2THF was isolated and its structure determined (Edema *et al.* 1989). In the majority of those complexes of metals other than vanadium which have been characterised the mode of bonding of the dinitrogen is end-on terminal. Only in a few instances can the ligated N_2 be reduced to NH_3, but *bis* N_2 complexes of Mo and W of the type trans-[Mo(N$_2$)$_2$-(dppe)$_2$] (dppe = Ph$_2$PCH$_2$CH$_2$PPh$_2$) are reactive at ambient temperature and pressure, and have been extensively studied. In a sequence of reactions which can be cycled chemically or electrochemically (see Pickett *et al.* 1988) the most stable and easily observable intermediate in the reduction of N_2 to NH_3 or N_2H_4 is M = N-NH$_2$ (where M is Mo or W). This entity in acid or base reacts to form free N_2H_4 and provides a parallel with the dinitrogen hydride-enzyme bound intermediate detectable in enzymic N_2 reduction. M = N-NH$_3{}^+$ has now been observed and characterised, it is stable for M = W, but reacts to form NH_3 when

$M = Mo$ (see Richards 1988). In the chemical cycle protonation of N_2 occurs in a stepwise manner with electrons flowing from the metal atom as protons are picked up from solution (Scheme 3). The modelling based on Mo and analogue chemistry provided the background against which the scheme for the mechanism of reduction of N_2 by Mo-nitrogenase was formulated. In the case of vanadium, this chemistry has yet to be developed.

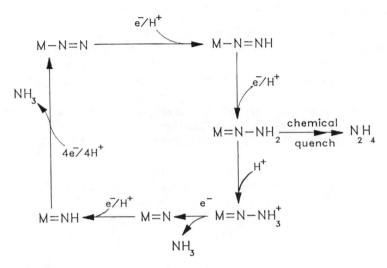

SCHEME 3

The reduction of N_2 to NH_3 in metal complexes. In this sequence N_2 is reduced by protons at the terminal nitrogen and electrons are supplied by the metal. If electrons are supplied from an electrode the reaction scheme can be cycled (See Pickett *et al.* 1988; Newton 1988; Leigh 1988; Richards 1988, for review).

In contrast, two of the best known aqueous systems which reduce N_2 are based on vanadium. The V^{III}/V^{II} system is a strong redox couple (Em -255mV) and in the presence of Mg^{2+} the heterogeneous V^{II} Mg^{II} alkaline gel catalyses the stoichiometric reduction of N_2 to N_2H_4 and NH_3 (Denisov *et al.* 1977). It has been proposed that in this system N_2 binds in a bridging mode to 2V atoms and that free N_2H_4 occurs as an intermediate in the route to NH_3 formation:

$$2V_2^{2+} + N_2 \xrightarrow{\quad\quad} V_2^{2+}N_2V_2^{2+} \xrightarrow{\quad 4H^+ \quad} 4V^3 + \underset{\underset{2NH_3}{\overset{|}{V^{II}}}}{N_2H_4} \qquad (4)$$

The reduction of N_2 occurs at room temperature and pressure, and the ratio of N_2H_4 and NH_3 as reduction products depends on pN_2 and the V/Mg ratio. The reduction is inhibited by CO and in ^2H_2O [cis $^2H_2$] ethylene is the product of C_2H_2 reduction; both of these properties are shared with Mo and V-nitrogenases.

Dinitrogen is also reduced by the complex formed by V^{II} and catechol at alkaline pH values (Nikonova *et al.* 1972). This system shows strong parallels with the reactions catalysed by nitrogenase and provides a good analogue. In the absence of added N_2, protons are reduced to H_2, under N_2 the evolution of H_2 is inhibited, but not completely and the limiting stoichiometry is:

$$8V^{2+} + 8H^+ + N_2 \longrightarrow 2NH_3 + H_2 + 8V^{3+} \qquad (5)$$

This reaction is similar in the ratio of $2NH_3 : H_2$ to that of Mo-nitrogenase functioning under optimum conditions. As with nitrogenase, N_2H_4 is not a reduction product, but if this system is chemically quenched with acid a small amount of N_2H_4 is detected and is presumed to arise from an intermediate in N_2 reduction.

The stereochemistry of C_2H_2 reduction is the same as Mo and V-nitrogenase with [cis $^2H_2$] ethylene as a product. This system also forms some C_2H_6 as a reduction product of C_2H_2, a property characteristic of the VFe protein of V-nitrogenase.

The aqueous chemical systems described above are difficult to characterise and mechanistic conclusions are often controversial but it is clear that N_2 reduction by V^{2+} has highly specific steric and pH requirements.

Outlook

The similarity between the Mo and V-nitrogenase systems suggests that they are derived from a common ancestral nitrogenase system. Although the availability of Mo to the organism is an important factor in determining which system is synthesised, it is too early to be sure that other environmental factors do not have a role in regulating their expression. This, and the unknown distribution of V-nitrogenase in Nature makes the assessment of the contribution of the V system to the nitrogen cycle difficult. It has been evident for many years that metals other than Mo can activate dinitrogen in chemical systems leading to the formation of NH_3. The discovery of V-nitrogenase has provided a new stimulus for investigation of the chemistry of such systems and will provide an important basis for structural and mechanistic studies of Mo-independent nitrogenases.

Acknowledgements

I thank the many coworkers who have made significant contributions to the work on V-nitrogenase of *A. chroococcum* discussed above, in particular Professors Barry Smith and Mike Dilworth and Drs. David Lowe, Roger Thorneley, Dick Miller, Rob Robson and Richard Pau.

References

Arber, J.M.; Dobson, B.R.; Eady, R.R.; Stevens, P.; Hasnain, S.S.; Garner, C.D. and Smith, B.E. 1987, Nature (Lond.) *325*, 372-374.

Arber, J.M.; Flood, A.C.; Garner, C.D.; Gormal, C.A.; Hasnain, S.S. and Smith, B.E. 1988, Biochem. J. *252*, 421-425.

Arber, J.M.; Dobson, B.R.; Eady, R.R.; Hasnain, S.S.; Garner, C.D.; Matsushita, T.; Nomura, M. and Smith, B.E. 1989, Biochem. J. *258*, 733-737.

Ashby, G.A.; Dilworth, M.J. and Thorneley, R.N.F. 1987, Biochem. J. *247*, 547-554.

Benneman, J.R.; McKenna, C.E.; Lie, R.F.; Traylor, T.G.; Kamen, M.D. 1972, Biochim. Biophys. Acta *264*, 25-28.

Bergström, J.; Eady, R.R. and Thorneley, R.N.F. 1988, Biochem. J. *251*, 165-169.

Bishop, P.E.; Jarlenski, P.M.L. and Heatherington, D.R. 1980, Proc. Natl. Acad. Sci. USA *77*, 7342-7346.

Bishop, P.E.; Jarlenski, D.M. and Heatherington, D.R. 1982, J. Bacteriol. *150*, 1244-1251.

Bishop, P.E.; Premakumar, R.; Dean, D.R.; Jacobson, M.R.; Chisnell, J.R.; Rizzo, T.M. and Kopczynski, J. 1986a, Science *232*, 92-94.

Bishop, P.E.; Hawkins, M.E. and Eady, R.R. 1986b, Biochem. J. *238*, 437-442.

Bortels, H. 1930, Archiv für Mikrobiologie *1*, 333-342.

Bortels, H. 1936, Zentralbl. Bakteriol. Parasitenkd. Infektionskr. Abt 2 *95*, 193-218.

Burgess, B.K. 1985, In Nitrogen Fixation Research Progress (Evans, H.J.; Bottomley, P.J. and Newton, W.E. eds.) pp 543-549, Dardrecht: Martinus Nijhoff.

Burns, R.C.; Fuchman, W.H. and Hardy, R.W.F. 1971, Biochim. biophys. Res. Commun. *42*, 353-358.

Chen, C.K.K.; Chen, J.S. and Johnson, J.L. 1986, J. Bacteriol *166*, 162-172.

Chisnell, J.R.; Premakumar, R. and Bishop, P.E. 1988, J. Bacteriol. *170*, 27-33.

Denisov, N.T.; Shuvalova, N.I. and Shilov, A.E. 1977, Kinet Katal *18*, 1606-1610.

Dilworth, M.J.; Eady, R.R.; Robson, R.L. and Miller, R.W. 1987, Nature (Lond.) *327*, 167-168.

Dilworth, M.J.; Eady, R.R. and Eldridge, M. 1988, Biochem. J. *249*, 745-751.

Dixon, R. 1989, in The Nitrogen and Sulphur Cycles (Cole, J.A. and Ferguson, S.J. eds.) pp 417-438, Cambridge University Press, Cambridge.

Eady, R.R. 1986, in Nitrogen Fixation Vol 4 (Broughton, W.J. and Pühler, A. eds.) pp 1-49, Oxford University Press, Oxford.

Eady, R.R.; Smith, B.E.; Cook, K.A. and Postgate, J.R. 1972, Biochem, J. *128*, 655-675.

Eady, R.R.; Robson, R.L.; Richardson, T.H.; Miller, R.W. and Hawkins, M. 1987, Biochem. J. *244*, 197-207.

Eady, R.R.; Richardson, T.H. Miller, R.W.; Hawkins, M. and Lowe, D.J. 1988, Biochem. J. *256*, 189-196.

Edema, J.J.H.; Meetsma, A. and Gamborotta, S. 1989, J. Am. Chem. Soc. *111*, 6878-6880.

Eidsness, M.K.; Frank, A.M.; Smith, B.E.; Flood, A.C.; Garner, C.D. and Cramer, S.P. 1986, J. Amer. Chem. Soc. *108*, 2746-2747.

Evans, D.; Jones, R,; Woodley, P. and Robson, R. 1988, J. Gen. Microbiol *134*, 931-942.

Filler, W.A.; Kemp, R.M.; Ng, J.C.; Hawkes, T.R.; Dixon, R. and Smith, B.E. 1986, Eur. J. Biochem *160*, 371-377.

Garner, C.D.; Arber, J.M.; Harvey, I.; Hasnain, S.S.; Eady, R.R.; Smith, B.E.; de Boer, E. and Wever, R. 1989, Polyhedron. *8*, 1649-1652.

George, G.N.; Coyle, C.L.; Hales, B.J. and Cramer, S.P. 1988, J. Am. Chem. Soc. *110*, 4057-4059.

Hagen, W.R.; Eady, R.R., Dunham, W.R. and Haaker, H. 1985, FEBS. Lett. *189*, 250-254.

Hales, B.J.; Case, E.E.; Morningstar, J.E.; Dzeda, M.F. and Mauterer, L.A. 1986a, Biochemistry *25*, 7251-7255.

Hales, B.J.; Langosch, D.J. and Case, E.E. 1986b, J. Biol. Chem. *261*, 15301-15306.

Hales, B.J.; True, A.E. and Hoffman, B.M. 1989, J. Amer. Chem. Soc. *111*, 8519-8520.

Hill, S. 1988, FEMS Microbiol. Rev. *54*, 111-130.

Hoover, T.R.; Imperial, J.; Ludden, P.W. and Shah, V.K. 1988, BioFactors *1*, 199-205.

Howard, T.B.; Davis, R,; Moldenhaver, B.; Cash, V.L. and Dean, D. 1989, J. Biol. Chem. *264*, 11270-11274.

Hughes, D.L.; Pomberiro, A.J.L. and Richards, R.L. 1988, J. Chem. Soc. Chem. Comm., 1052-1053.

Jacobson, M.R.; Brigle, K.E.; Bennet, L.T.; Setterquist, R.A.; Wilson, M.S.; Cash, V.L.; Beynon, J.; Newton, W.E. and Dean, D.R. 1989, J. Bacteriol *171*, 1017-1027.

Joerger, R.D.; Jacobson, M.R.; Premakumar, R.; Wolfinger, E.D. and Bishop, P.E. 1989a, J. Bacteriol. *171*, 1075-1086.

Joerger, R.D.; Jacobson, M.R. and Bishop, P.E. 1989b, J. Bacteriol *171*, 3258-3267.

Joerger, R.; Loveless, T.; Pau, R.; Michenall, L. and Bishop, P.E. 1990, J. Bacteriol, submitted.

Kennedy, C.K.; Gamal, R.; Humphrey, R.; Ramos, J.; Brigle, K. and Dean. D. 1986, Mol. Gen. Genet. *205*, 318-325.

Kentemitch, T.; Danneberg, G.; Hundeshagen, B. and Bothe, H. 1988, FEMS Microbiol. Lett. *51*, 19-24.

Kovacs, J.A. and Holm, R.H. 1986, J. Am. Chem. Soc. *108*, 340-341.

Lindahl, P.A.; Teo, B.K. and Orme-Johnson, W.H. 1987, Inorg. Chem. *26*, 3912-3916.

Leigh, G.J. 1988, J. Molec. Cat. *47*, 363-379.

Lowe, D.J.; Thorneley, R.N.F. and Smith, B.E. 1985, In Metallo Proteins Part I (Harrison, P.M. ed.) pp 207-249, London: Macmillan Press.

Luque, F and Pau, R.N. 1990, J. Bacteriol, in press.

Meyer, J.; Gaillard, J. and Moulis, J.M. 1988, Biochemistry *27*, 6150-6156.

McKenna, C.E.; Benneman, J.R. and Traylor, T.C. 1970, Biochim. Biophys. Res. Commun. *41*, 1501-1508.

Merrick, M.J. 1988, In Nitrogen Fixation: Hundred Years After (Bothe, H.; de Bruijn, F.J. and Newton, W.E. eds.) pp 293-302, G. Fischer, Stuttgart.

Miller, R.W. and Eady, R.R. 1988a, Biochem. J. *256*, 429-432.

Miller, R.R. and Eady, R.R. 1988b, Biochim. biophys. Acta. *952*, 290-296.

Morningstar, J.E.; Johnson, M.K.; Case, E.E. and Hales, B.J. 1987, Biochemistry *26*, 1795-1800.

Morningstar, J.E. and Hales, B.J. 1987, J. Am. Chem. Soc. *109*, 6854-6855.

Newton, W.E.; Burgess, B.K.; Cummings, S.C.; Lough, S.; McDonald, J.W.; Robinson, J.F.; Conradson, S.D. and Hodgson, K.O. 1984, in Advances in Nitrogen Fixation Res. (Veeger, C. and Newton, W.E. eds.) pp 160-167, Nijhoff/Junk, The Hague.

Newton, W.E. 1988, in Encylopedia of Chemical Technology Vol. 15., pp 942-968, J. Wiley & Sons, New York.

Nikonova, L.A.; Ovcharenko, A.G.; Efimov, O.N.; Avilov, V.A. and Shilov, A.E. 1972, Kinet Katal *13*, 1602-1605.

Orme-Johnson, W.H. 1985, Ann. Rev. Biophys. Chem. *14*, 419-459.

Page, W.J. and Collinson, S.K. 1982, Con. J. Microbiol. *28*, 1173-1180.

Pau, R.N. 1989, TIBS *14*, 183-186.

Pau, R.N.; Mitchenall, L.A. and Robson, R.L. 1989, J. Bacteriol. *171*, 124-129.

Paul, N. and Merrick, M. 1989, Eur. J. Biochem. *178*, 675-682.

Pickett, C.J.; Cate, K.; MacDonald, C.J.; Mohammed, M.Y.; Ryder, K.S. and Tarmin, J. 1988, In Nitrogen Fixation: Hundred Years After, (Bothe, H.; de Bruijn, F.J. and Newton, W.E. eds.) pp 51-55, Gustav Fischer, New York.

Premakumar, R.; Lemos, E.M. and Bishop, P.E. 1984, Biochim. biophys. Acta. *797*, 64-70.

Pretorius, I.M.; Rawlings, D.E.; O'Neill, E.G.; Jones, W.A.; Kirby, R. and Woods, D.R. 1987, J. Bacteriol. *169*, 367-370.

Robinson, A.C.; Dean, D.R. and Burgess, B.K. 1987, J. Biol. Chem. *262*, 14327-14332.

Robson, R.L.; Eady, R.R.; Richardson, T.J.; Miller, R.W.; Hawkins, M. and Postgate, J.R. 1986a, Nature (Lond.) *322*, 388-390.

Robson, R.L. 1986, Arch. Microbiol. *146*, 74078.

Robson, R.L.; Woodley, P.R.; Pau, R.N. and Eady, R.R. 1989, EMBO J. *8*, 1217-1224.

Raina, R.; Reddy, M.A.; Ghosal, D. and Das, H.K. 1988, Mol. Gen. Genet. *214*, 121-127.

Richards, R.L. 1988, Chemistry in Britain *136*, 133-135.

Santero, E.; Toukdarian, A.; Humphrey, R. and Kennedy, C. 1988, Molec. Microbiol. *2*, 303-314.

Shah, V.K.; Vgalde, R.A.; Imperial, J. and Brill, W.J. 1984, Am. Rev. Biochem. *53*, 231-257.

Smith, B.E.; Campbell, R.; Eady, R.R.; Eldridge, M.; Ford, C.M.; Hill, S.; Kavanagh, E.P.; Lowe, D.J.; Miller, R.W.; Richardson, T.H.; Robson, R.L.; Thorneley, R.N.F. and Yates, M.G. 1987, Phil. Trans. R. Soc. Lond. B. *317*, 131-146.

Smith, B.E.; Eady, R.R.; Lowe, D.J. and Gormal, C. 1988, Biochem. J. *250*, 299-302.

Souillard, N.; Magot, M.; Possot, O. and Sibold, L. 1988, J. Mol. Evol. *27*, 65-76.

Stephens, P.J. 1985, in Molybdenum Enzymes (Spiro, T.G. ed.) pp 117-159, Wiley & Sons, New York.

Thorneley, R.N.F. and Lowe, D.J. 1985, in Molybdenum Enzymes (Spiro, T.G. ed.) pp 221-284, Wiley & Sons, New York.

Thorneley, R.N.F.; Bergström, N.H.J.; Eady, R.R. and Lowe, D.J. 1989, Biochem. J. *257*, 789-794.

VII. Insulin Mimetic Effects of Vanadium

YORAM SHECHTER, JOSEPH MEYEROVITCH, ZVI FARFEL,
JOSEPH SACK, RAFAEL BRUCK, SHIMON BAR-MEIR, SHIMON
AMIR, HADASSA DEGANI and STEVEN J.D. KARLISH
Department of Hormone Research (Y.S.), Biochemistry (S.J.D.K.), Isotopes (H.D.; S.A.), Weizmann Institute of Science, Rehovot 76100, Department of Pediatrics (J.M.;J.S.), Internal Medicine (Z.F.), Sheba Medical Center, Tel Hashomer, and Department of Gastroenterology (R.F.; S.B.M.), Wolfson Medical Center, Holon, Israel

Historical Perspective

As can be seen in the previous chapters, vanadium attracted the attention of many scientists from various disciplines for three decades. The interest in vanadium increased when it became apparent that vanadium is an endogenous component present in trace amounts in tissues of higher animals, and is essential for growth and development, as well as for the normal growth of mammalian cells in culture (reviewed by Post *et al.*, 1979; Simons, 1979; Nechay *et al.*, 1984). In 1980 a new interest in vanadium emerged when it was found that like insulin, vanadate increases hexose uptake and glucose metabolism in isolated rat adipocytes (Shechter and Karlish, 1980; Dubyak and Kleinzeller, 1980). Shortly after, vanadate was shown to mimic insulin in inhibiting lipolysis (Degani *et al.*, 1981). In the next several years vanadate was demonstrated to mimic virtually all or most of the documented actions of the hormone in *in vitro* systems. A new turning point occurred in 1985 when Heyliger, McNeill and coworkers demonstrated that vanadate administered, *orally*, to diabetic hyperglycemic rats lower their blood glucose levels to normal values. This opened new avenues for investigators in the fields of metabolic as well as diabetes research.

Insulin Action

Before reviewing the insulin mimetic effects of vanadium a brief outline on the current state of insulin action is necessary. Insulin elicits a remarkable array of biological responses. The physiologically important target tissues for insulin with respect to glucose homeostasis are liver, muscle and fat, but insulin exerts potent regulatory effects on other cell types as well. In attempting to understand the actions of insulin one should bear in mind that insulin is the primary hormone responsible for signalling the storage and utilization of basic nutrients. Thus, in general, insulin activates the transport systems and the enzymes involved in intracellular utilization and storage of glucose, amino acids, and fatty acids, while inhibiting the catabolic processes evoked by counter-regulatory hormones such as

N. Dennis Chasteen (ed.), Vanadium in Biological Systems, 129–142.
© *1990 Kluwer Academic Publishers. Printed in the Netherlands.*

breakdown of glycogen, fat and proteins. Insulin is perhaps the only anabolic hormone that also arrests catabolic processes.

Traditionally, the actions of insulin are classified into three groups based on their kinetics. The *immediate or the rapid effects* of insulin occur within seconds or minutes after the addition of the hormone, and include activation of glucose and ion transport systems and the covalent modification (i.e., phosphorylation and dephosphorylation) of pre-existing enzymes. The *intermediate effects*, such as the induction of ornithine decarboxylase and tyrosine amino transferase activity, occurs 3 to 6 hours after the addition of insulin and involve induction of genes and expression of proteins. The *long term* effects of insulin require many hours to several days. This category includes insulin's effects to stimulate DNA synthesis, cell proliferation and cell differentiation. The rapid, intermediate and long term actions of insulin may not share a single common mechanistic pathway, but may be the result of diverging and converging pathways mediated by the insulin and IGF receptors.

The first step in insulin action is the high affinity-specific binding ($K_d \cong 3$ nM) of the hormone to externally located cell surface receptor sites. Low receptor occupancy ($\sim 2\%$ at the total surface receptor sites) is usually sufficient to elicit the maximal biological responses to insulin (Reviewed by Shechter, 1985). The insulin receptor is a large transmembraneous glycoprotein of approximately 300 kDa and composed of two extracellular α-subunits (135 kDa each) and two transmembrane β-subunits (95 kDa each) linked by disulfide bands to form β-α-α-β heterotetramer (Czech, 1984; Kahn and White, 1988).

In 1982 it was reported that the insulin receptor is an insulin-activated tyrosine-specific protein kinase (Kasuga *et al.*, 1982). This finding was particularly intriguing because the amount of phosphotyrosine in cells is $<0.1\%$ of the combined amount of phosphoserine and phosphothreonine. Insulin binding to the extracellular α-subunits of the heterotetrameric molecule, leads to a rapid intra-molecular autophosphorylation of the β-subunits. This occurs as a result of an increase in the V_{max} (rather than the K_m) for ATP. Autophosphorylation requires the presence of Mn^{++} and occurs at several tyrosine residues (White *et al.*, 1988). This phosphorylation is autocatalytic and activates the tyrosine kinase activity toward other substrates many fold. Autophosphorylation and kinase activation occurs in cells at physiological concentrations of insulin and is maximal within seconds, indicating that it may be sufficient to stimulate the early actions of insulin.

In intact cells, the insulin receptor is also phosphorylated on serine and threonine residues. This occurs by the action of other cellular kinases.

In contrast to tyrosine phosphorylation, which is activating, serine and threonine phosphorylation is inactivating (Takayama *et al.*, 1988). Thus, the insulin receptor is regulated in its activity by multiple phosphorylation. Ultimately, many of the enzymes that are acutely modulated by insulin action undergo either Thr/Ser phosphorylation (e.g., acetyl-Co A carboxylase, ATP-citrate lyase, S-6-ribosomal protein) or Thr/Ser dephosphorylation (e.g., glycogen synthase, pyruvate dehydrogenase, hormone-sensitive lipase). The steps in the cascades which link between tyrosine phosphorylation and the Ser/Thr phosphorylation and dephosphorylation cascades are still lacking.

After insulin binding, the occupied receptors are aggregated along the plasma membrane and then internalized rapidly. The bound insulin is mainly degraded and the receptor is recycled back to the cell surface. Thus the early events at the level of the receptor following insulin binding seems to be a conformational change at the receptor; intramolecular autophosphorylation: receptor redistribution, and receptor internalization. All these events may be essential and are rapid enough to account for the early actions of insulin.

Glucose Transport Regulation

The effect of insulin in stimulating glucose uptake into muscle and adipose tissues forms a central point in its physiological role. This effect of insulin on glucose influx can be assessed independently of the effect of the hormone on glucose metabolism by using non-metabolizable glucose derivatives, e.g. 3-0-methyl glucose (Czech, 1980). Insulin increases the rate of 3-0-methyl glucose uptake 20 to 30 fold. The effect depends on an increase in the maximal rate of transport (V_{max}) with little or no effect on the K_m. The K_m of 3-0-methyl glucose transport in adipocytes is in the low physiologic range (2-5 mM). Transport is stereospecific, and occurs after a short lag period (20-45 seconds at 37°C) following insulin stimulation. Activation and reversal of this process are independent of protein synthesis, but require endogenous ATP. The kinetic behavior is consistent with an insulin-dependent increase in number of hexose transporters in the plasma membrane. Insulin does not stimulate glucose uptake in liver and several other tissues, despite the sensitivity of these tissues to other insulin effects. This suggests that there are several types of glucose transporters which may differ in the molecular regulation.

One major component in the mechanism of insulin stimulation of glucose transport is a temperature and energy dependent translocation of intracellular vesicles containing the glucose transporter (GT) to the plasma membrane (Suzuki and Kono, 1980; Karnieli et al., 1981; Kono et al., 1982; Simpson and Cushman, 1986; James et al., 1988). The effect is reversible, and on removing insulin, the glucose transporters return to the intracellular pool. There remains considerable debate, however, as to whether translocation accounts for the entire insulin stimulating effect, since in most studies there is a 3 to 10-fold increase in membrane transporters due to translocation, but a 20 to 30 fold increase in V_{max} for transport. In addition, translocation has been difficult to demonstrate in muscle, despite the clear effect of insulin to stimulate transport in this tissue. This observation has lead to the suggestion that there is an activation (or increase in the intrinsic activity) of the plasma membrane glucose transporters in addition to the translocation. The notion of activation is also supported by the observation that some adrenergic agents and activators of protein kinase C, also stimulate transport, but have little or no effect on translocation (reviewed in Simpson and Cushman, 1986).

Chemical and Biochemical Properties of Vanadium

As mentioned above, vanadium is now recognized as an essential nutritional element in higher animals but its function remains unclear (Macara, 1980; Ramasarma and Crane, 1981). The chemistry of vanadium is complex since it can exist in a variety of oxidation states ranging from -1 to +5, and can be found in a multitude of polymeric forms. At physiological concentrations in mammals and avians, free vanadium, exists in a hydrated monomeric state. In body fluids at pH 4-8, the predominant species found is orthovanadate (HVO_4^{2-} and $H_2VO_4^-$) which exists in the +5 oxidation state (Simons, 1979; Nechay, 1984; Ramasarma and Crane, 1981; Macara, 1980). Most tissues contain intracellular vanadium at a concentration of about 0.1-1.0 μM (Simons, 1979). The probable mode of entry of VO_3^- into the cell is by means of an anion carrier system and, once inside it is reduced non enzymatically, possibly by glutathione, to vanadyl (VO^{2+}, +4 oxidation state). VO_4^{3-} is a structural analog of orthophosphate (Macara, 1980), and this may account for its inhibitory effect on enzymes involved in phosphate transfer and release mechanisms (Simons, 1979; Macara, 1980; Lopez et al., 1976; Lindquist et al., 1976). Vanadate can easily adopt a stable trigonal bipyramidal structure which resembles the transition state of phosphate during reaction (Macara, 1980). The list of enzymes that are inhibited by vanadate includes ATP phosphohydrolases, ribonuclease, glyceraldehyde 3-phosphate dehydrogenase, alkaline phosphatase, Ca^{2+}/Mg^{2+} ATPase, Na^+/K^+ ATPase, and phosphotyrosyl protein phosphatase (reviewed by Nechay, 1984).

Perhaps the best described effect of vanadate is its inhibitory action on Na^+/K^+ ATPase activity. It appears that vanadate binds to both the low and high affinity binding sites for ATP of the enzyme. The high affinity site of VO_4^{3-} action ($K_{d,app}$ = $4 \times 10^{-9}M$) corresponds to the low affinity ATP binding site, and vice versa. VO_4^{3-} binding to the E_2 (K^+-dependent) state of the enzyme results in stabilization of the E_2 state and thereby prevents transition towards the E_1 (Na^+-dependent) state. It has been shown that Mg^{2+}, K^+, ouabain, and dimethyl sulfoxide favor the formation of the E_2 state, resulting in increased vanadate binding. Conversely, ATP, Na^+, and oligomycin, which favor formation of the E_1 state, result in a decrease in VO_4^{3-} binding. A physiological role for VO_4^{3-} regulation of Na^+ pump activity has been postulated. Vanadate, however, should interact with the cytoplasmic surface of the Na^+/K^+ ATPase (Cantley et al., 1978). Intracellular vanadium is found primarily in the form of VO^{2+}, which is a relatively ineffective inhibitor of Na^+/K^+ ATPase in vitro. Indeed, as will be discussed later, exogenously-added vanadate, up to a concentration of 1 mM, does not inhibit the Na^+/K^+ ATPase in a wide variety of mammalian cells (reviewed by Nechay, 1984). An exception is the pancreatic-islet-beta cell which secretes insulin. In this cell type (which differs from other mammalian cells in many respects) vanadate inhibits Na^+/K^+ ATPase (K_i = 1-10 μM) and therefore induces insulin secretion (Fagin et al., 1987). This may be an additional contributing factor in maintaining euglycemia in diabetic animal models (later paragraph).

In vitro Actions of Vanadate

Isolated, intact cells are relatively less complex experimental systems for analyzing the actions of insulin. The observations made in cells can later be studied in more complex systems such as perfused liver or muscle and finally applied to intact experimental animal models (i.e. to mice and rats). The insulin-like effects of vanadium were initially observed in isolated rat adipocytes. Exogenously added metavanadate mimicked insulin in stimulating hexose uptake (Dubyak and Kleinzeller, 1980), glucose oxidation (Shechter and Karlish, 1980), lipogenesis (Shechter and Ron, 1986), and the inhibition of lipolysis (Degani *et al.*, 1981). The half-maximally effective concentrations of VO_3^- or VO_4^{3-} for mediating the various insulin-like effects, range between 0.05 and 0.2 mM. As mentioned earlier, these effects do not seem to be secondary to the inhibition of Na^+/K^+ ATPase activity. This is because VO_3^- over a concentration range of 0.5-1 mM does not inhibit Na^+/K^+ ATPase activity at all in *intact* adipocytes (Dubyak and Kleinzeller, 1980; Shechter and Karlish, 1980).

The fate of the added VO_3^-, after entering the intact adipocyte, was evaluated using electron spin resonance spectroscopy (Degani *et al.*, 1981). This study revealed that vanadate is reduced intracellularly to VO^{2+} and that in the reduced state it was found to complex exclusively to reduced glutathione (GSH). The major vanadyl complex found has a VO^{2+} : GSH stoichiometry of 1 : 1. At least two such VO^{2+}-GSH complexes could be detected, whereas no complexes of VO^{2+} to other ligands (i.e. to ATP) could be observed. It is conceivable that VO^{2+} binding to GSH prevents the reoxidation to VO_4^{3-}, which would tend to occur at the intracellular pH. This, together with the lack of inhibition of Na^+/K^+ ATPase, suggests, that intracellular VO^{2+} (rather than VO_3^-) alone or complexed to reduced glutathione is the relevant chemical species responsible for mimicking the actions of insulin.

Further studies have indicated that insulin and vanadate share several common features. Vanadate at sufficiently high concentrations, maximally stimulates hexose uptake, glucose oxidation, and lipogenesis. Furthermore, no increment in stimulation could be achieved by the addition of insulin to vanadate-stimulated cells, and vice versa (unpublished data). In addition, both agents show the same concentration dependency on extracellular glucose. Agents or conditions which suppress the effects of insulin, such as anti-calmodulin drugs, polymyxin B, bicarbonate-depleted buffers, and exogenously-added ATP are equipotent in suppressing vanadate-mediated effects. Also, similar rates in the termination of lipogenesis ($t_{1/2}$ = 14 ± 3 min) were observed after removing either insulin or vanadate from stimulated adipocytes (reviewed by Shechter *et al.*, 1988).

In recent years additional insulinomimetic effects of vanadium have been documented in other cell types as well (summarized in Table I). Thus, vanadate also stimulates hexose uptake in skeletal muscle; activates glycogen synthase in adipose tissue, liver and muscle; enhances K^+ uptake in cardiac muscle cells; exerts Ca^{2+} influx in adipose tissue; inhibits Ca^{2+}/Mg^{2+} ATPase in rat adipocytic plasma membranes; elevates intracellular pH in A-431 cells, and suppresses the

Table I. Documented insulin-like actions of vanadium in *in vitro* systems.

Activity reference	Direction of activation	Target tisue
Hexose transport Dubyak and Kleinzeller, 1980	Stimulated	Rat adipocytes
Hexose transport Clark *et al.*, 1985	Stimulated	Skeletal muscle
Lipogenesis Shechter and Ron, 1986	Stimulated	Rat adipocytes
Glucose oxidation Shechter and Karlish, 1980 Dubyak and Kleinzeller, 1980 Duckworth *et al.*, 1988	Stimulated	Rat adipocytes
Glucose oxidation Clark *et al.*, 1985	Stimulated	Skeletal muscle
Lipolysis Degani *et al.*, 1981 Duckworth *et al.*, 1988	Inhibited	Rat adipocytes
Glycogen synthase Tamura *et al.*, 1984	Stimulated	Skeletal muscle
Tamura *et al.*, 1984 Jackson *et al.*, 1988		Rat adipocytes Rat hepatocytes
Secretion of Jackson *et al.*, 1988 apolipopro-tein B	Inhibited	Rat hepatocytes
Mitogenic activity Hori and Oka, 1980 Smith, 1983 Canalis, 1983 Reid and Reid, 1987	Augmented	Various cultured cells
Translocation of Kadota *et al.*, 1986 IGF-II receptors	Stimulated or augmented	Rat adipocytes
Potassium uptake Werdan *et al.*, 1982	Stimulated	Cardiac muscle cells
Ca^{2+}/Mg^{2+} ATPase Delfert and McDonald, 1985	Inhibited	Plasma membranes from rat adipocytes
Ca^{2+} influx Clausen *et al.*, 1981	Stimulated	Adipose tissue
Intracellular pH Cassel *et al.*, 1984	Elevated	A-431 cells

secretion of apolipoprotein B from rat hepatocytes (Table I). Unlike the effects of insulin in certain cells, the mitogenic related delayed events of insulin (or EGF) are augmented (Table I) in the presence of vanadate, but are not stimulated by vanadate alone. These delayed effects of insulin differ from the rapid effects also in the sense that insulin must be constantly bound to its cellular receptor for many hours. Hence, intracellular signal(s) for DNA synthesis are transitory and disappear on removal of insulin. Vanadate added to the cells stabilizes the messenger(s) in such a way that the signal does not decrease (Reid and Reid, 1987).

Recently, Duckworth *et al.* (1988) observed that vanadate was more potent than insulin in stimulating glucose oxidation ($^{14}CO_2$ production) from $[1\text{-}^{14}C]$glucose in rat adipocytes. Glucose oxidation from $[6\text{-}^{14}C]$glucose was identically stimulated by either insulin or vanadate. $^{14}CO_2$ production from $[1\text{-}^{14}C]$glucose is a measurement of pentose phosphate shunt activity, whereas $^{14}CO_2$ production from $[6\text{-}^{14}C]$glucose reflects glycolytic flux. Therefore, vanadate may have a larger effect than insulin on the cellular pentose phosphate shunt activity.

Effect of Vanadate in Isolated Perfused Liver

Perfused isolated liver is a suitable system for studying the action of insulin and it makes the junction point between the relatively simpler *in vitro* systems (isolated or broken cell preparations) and the more sophisticated and complexed intact animal model. *In vivo*, secreted insulin reaches first the liver via the portal vein and exerts a multitude of bioactions in this organ (reviewed by Kahn and Shechter, 1990). The isolated – perfused liver responds to physiologically added concentrations of insulin as well (Mondon *et al.*, 1974). For example, insulin suppresses the release of glucose, potassium and amino acids from perfused liver (Mondon *et al.*, 1974). The release of glucose is predominantly inhibited due to the arrest of biochemical pathways which leads to glycogen breakdown.

We have recently observed that vanadate at very low concentrations (IC_{50} = 0.5-1 μM) inhibits effectively glucose release from isolated-perfused rat liver (Bruck *et al.*, submitted manuscript). The magnitude of the effect was larger than that obtained by insulin. Several other parameters such as the rate of initiation, and the rate of termination of the effect resembled those of the hormone (submitted, and in preparation). Thus, this physiologically important hepatic function of insulin is also mimicked by vanadate.

Studies in Intact Experimental Animals

The high dose streptozotocin-treated rats (ST-rats) is a suitable experimental diabetic model as it reflects symptoms of both Type I and Type II diabetes. These rats exhibit low endogenous production of insulin and high levels of circulating glucose. The high levels of glucose (hyperglycemia) induce tissue alterations which lead to poor responsiveness of liver, muscle and fat to insulin (Kasuga *et al.*, 1978; Kobayashi and Olefsky, 1979). Also, the levels of key enzymes of carbohydrate metabolism and of liver glycogen are greatly diminished (Weinhouse, 1976; Gil *et al.*, 1986). These alterations in ST-rats and in diabetic patients are largely corrected by subcutaneous injections of insulin (Kasuga *et al.*, 1978; Kabayashi and Olefsky, 1979). Like other proteins, insulin is not adsorbed (therefore ineffective) on oral administration in mammals (Hirsova and Koldovsky, 1969). Vanadate, however, is a low molecular-weight, phosphate analog, and as such it is adsorbed orally, and reaches the circulation (Heyliger *et al.*, 1985; Meyerovitch *et al.*, 1987).

Heyliger *et al.* (1985) were the first to observe that the inclusion of sodium orthovanadate (Na_3VO_4; 0.8 mg/ml) and 80 mM NaCl in the drinking water of streptozotocin-treated diabetic rats alleviated some symptoms of diabetes. This treatment resulted in the normalization of blood glucose levels and the elimination of depressed cardiac performance. These effects were shown not to result from increased levels of endogenous insulin; therefore, insulin target tissues have been implicated as the site(s) of vanadate action. This study was further extended by Meyerovitch *et al.* (1987), who found that a lower concentration of vanadate

(NaVO$_3$, 0.2 mg/ml) in the drinking water was optimal for achieving stable normoglycemia in ST-rats over a period of several weeks. Blood glucose levels dropped to nearly normal values within 3-4 days after the inclusion of VO$_3^-$ in the drinking water (Meyerovitch et al., 1987; Blondel et al., 1989). Following the removal of VO$_3^-$ from the drinking water, normoglycemia persisted for another 3-4 days before the onset of hyperglycemia, indicating that vanadate-induced normoglycemia is reversible. ST-rats, which are known to be in a catabolic state, became anabolic (gaining 1.3 g/day) several days after receiving the lower vanadate dosage. The treatment also corrected various tissue alterations known to develop in ST-rats. For example, VO$_3^-$ therapy reduced the elevated insulin binding capacity of liver to normal capacity and partially restored the responsiveness of adipocytes to insulin.

It has been shown that vanadate therapy doubles the rate of 3-O-methyl glucose uptake in muscle and liver tissues in ST-rats. Thus, the normoglycemia observed in vanadate-treated ST-rats seems to result from VO$_3^-$ stimulation of glucose uptake and its metabolism in vivo, in agreement with the known action of VO$_3^-$ in in vitro systems.

No signs of VO$_3^-$ toxicity could be detected in ST-rats receiving lower doses of VO$_3^-$. The level of VO$_3^-$ in the serum of treated rats did not exceed 0.7 to 0.9 μg/ml, a concentration about 1/100 of that required to inhibit a variety of Na$^+$/K$^+$ ATPases of intact cellular systems.

Certain beneficial effects were observed in the liver of streptozotocin-treated rats upon treatment with VO$_3^-$ (Gil et al., 1988). As mentioned earlier, in ST-rats the level of several key enzymes involved in carbohydrate metabolism, as well as the level of glycogen is greatly diminished in ST rats (Weinhouse, 1976; Gil and Bartrons, 1986). Oral treatment with VO$_3^-$ for two weeks restored the level of glycogen from 40% to 109% of control rats, the level of glucokinase from 0% to 65% of control rats, and the level of the dephosphorylated (active) 6-phospho-fructo-2-kinase from 20% to 122% of healthy control rats. Thus, in addition to these profound effects this study further supports the notion that the tissue alterations of diabetes result from the state of hyperglycemia and are not the result of the low endogenous levels of insulin.

Central Effects of Vanadate

We have observed in control, healthy rats, receiving VO$_3^-$ orally, a decrease in food intake and correspondingly a significant reduction in body weight-gain. This metabolic change was transitory. The effect persisted for 4-5 days after which food intake and body weight-gain returned to normal value (Figure 1). Further studies have indicated that like in other tissues, vanadate also increases glucose uptake and glucose metabolism in brain tissue (Meyerovitch et al., 1989). Increased glucose metabolism within the CNS produces a signal to reduce eating. This behavioral effect was less consistently observed in diabetic rats. Possibly due to the apriory lower basal activity of glucose metabolism in the brain tissue of this diabetic model (Meyerovitch et al., 1989).

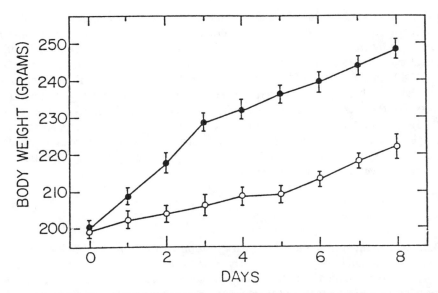

Fig. 1. Effect of orally administered vanadate on the daily weight gain of rats. Rats were supplied with water containing 80 mM NaCl ($\bullet$), or with water containing 80 mM NaCl and 0.2 mg/ml $NaVO_3$ ($\circ$). At the time points indicated in the figure, each rat in the group (n = 7) was weighed between 9 a.m. and 11 a.m. Each point represents the arithmetic mean of weight for 7 rats.

In adult mammals insulin may not reach the CNS and therefore is likely not to exert direct effects in this tissue. However, when insulin is injected centrally in experimental animals, it rapidly lowers the plasma glucose concentration. This effect occurs by blocking glucose output from the liver through the parasympathetic outflow (Szabo and Szabo, 1983). In contrast to insulin, central administration of vanadate in mice, produced systemic hyperglycemia (Amir et al., 1987). This effect was blocked by coadministration of 3-O-methyglucose, which is known to prevent glucose uptake and its metabolism. In summary, insulin acts within the CNS as a neurotransmitter. Vanadate does not mimic this action, but seems to increase glucose metabolism in brain, similar to its action in peripheral tissues.

Effects of Vanadate on Insulin Secretion

In addition to the documented actions of vanadate in peripheral tissues there is some scattered data concerning beneficial effects of vanadate on insulin secretion. *In vitro*, vanadate (~ 1 mM) induced biphasic secretion of insulin from incubated and perifused rat islets (Fagin et al., 1987). This effect may be secondary to the inhibition of Na^+/K^+ ATPase. As discussed earlier, the Na^+/K^+ ATPase of the intact β cell is exceptionally inhibited by vanadate (Fagin et al., 1987).

In ST-rats, vanadate treatment, and the concomitant normalization of blood glucose, was not accompanied by an increase in insulin secretion (Heyliger et al.,

1985). However, insulin secretion in the intact animal is under the multiple regulation of other pancreatic hormones, gut hormones and neurotransmitters (reviewed by Kahn and Shechter, 1990). Therefore, an improvement in the insulin secretory capacity of the islets may be unnoticed. In a recent study Pederson *et al.* (1989) have found that three weeks of vanadyl treatment of ST-rats following 13 weeks withdrawal yielded islets close to the size and insulin content of control islets, even though *in vivo* and *in vitro* insulin secretion was impaired (Pederson *et al.*, 1989). It is therefore conceivable that vanadyl treatment protects islets from further destruction in this diabetic model. The lack of improvement in insulin secretion may be explained by a hormonal and neuronal inhibitory regulation on the insulin-secreting pancreatic system.

Mechanism of Action of Vanadate

Data accumulated in the last few years from *in vivo* and *in vitro* studies seem to agree that vanadate bypasses the early receptor events of the insulin-dependent cascade (e.g. insulin binding, insulin-receptor-kinase (InsRK) autophosphorylation, and kinase activation). This is somewhat surprising because vanadate is a potent inhibitor of phosphotyrosine protein phosphatase (PTPase, Swarup, 1982) and after the tyrosine kinase activity of the insulin receptor was discovered, it has been assumed that cellular PTPase activity is responsible for terminating the effects of insulin by dephosphorylating the InsRK and other cellular intermediates. Vanadate alone, when incubated with cells, or isolated insulin receptors does not activate InsRK autophosphorylation nor does it increases to any detectable level cellular phosphotyrosine content of proteins (Kadota, 1986; 1987; Strout *et al.*, 1989; Shechter, unpublished observations). Thus, under basal conditions, this enzymatic activity is either low and/or in an inactive state[1]. In agreement with this, vanadate is equipotent in stimulating glucose metabolism in rat adipocytes in which about 50-60% of the InsRK were lost by down regulation, and are less responsive to insulin itself (Green, 1986).

   To summarize this issue, it seems that vanadate stimulates glucose uptake and glucose metabolism by either an alternative (insulin-independent) cascade mechanism, or at least by bypassing the early (receptor) events in the cascade. Either of these alternatives is of clinical interest, since *insulin resistance* in humans also includes defects in the insulin receptor itself (reviewed by Reddy and Kahn, 1988).

   Which alternative pathway(s) can lead to increased glucose transport activity (GTA)? We currently know that increased GTA is, at least in part, an exocytotic process in which internal glucose transporter-containing organelles are exocytosed to the plasma membrane. By analogy to other well-studied exocytotic processes in mammalian cells, four basic components are required for exocytosis to occur.

---

[1] Increased cellular phosphotyrosine content of proteins was observed by the combined treatment of cultured cells with vanadate and insulin, e.g. Bernier *et al.* (1988).

These are Ca^{2+} ions, a calcium binding protein, a fusogen, and a change in the intravesicular/cytoplasmic pH. Calcium ions and a Ca^{2+} binding protein are required to aggregate the vesicles in question with the plasma membrane, and a fusogen (i.e., cis-unsaturated fatty acid) facilitates the fusion step. Decreased intravesicular pH of the vesicles also assists in this process, but by a not well-understood mechanism. Fusion of chromaffin granules *in vitro* occurs with milli-molar concentrations of Ca^{2+} and synexin, but also a Ca^{2+} binding protein, which facilitates fusion at micromolar (physiological), concentrations of Ca^{2+}, was recently identified (calpactin, Drust and Creutz, 1988). Several lines of evidence indicate that a change in one (or more) of these four components leads to an increase or a decrease in GTA. For example, cis-unsaturated fatty acids promoted lipogenesis in rat adipocytes, as well as certain diacyl-glycerols that can be introduced into the cell interior and which may act as fusogens (Shechter and Henis, 1984; Stralfors, 1988). Inhibitors of calcium binding proteins, or chelation of intracellular Ca^{2+} suppresses increase rate of GTA (Shechter, 1984; Pershadsingh *et al.*, 1987). An increase in intracellular Ca^{2+} level in cells may not occur by insulin stimulus (Moore, 1983) but is positively associated with many of the pleiotropic effects of insulin in various tissues, including increased rates of glucose metabolism (reviewed by Czech, 1977).

Vanadate may enter into this scheme at several points. First, it may increase the level of cytoplasmic Ca^{2+} since both vanadate and vanadyl ions were shown to inhibit the high affinity Ca^{2+}/Mg^{2+} ATPase. Alternatively, it could alter both intracellular and intravesicular pH (Schindler and Shechter, in preparation). The effects of vanadate on intracellular pH have been previously documented. It should be noted that three different mechanisms which control intracellular pH are currently known in mammalian cells – two of which depend on the presence of bicarbonate in the external medium (reviewed by Thomas, 1989). We have observed that in rat adipocytes which are maintained in bicarbonate- and phos-phate-free buffer, the stimulating effect of vanadate on glucose metabolism is greatly abolished. The effect was restored by readding bicarbonate ions directly to the medium (Shechter and Ron, 1986a,b). Thus, a possible effect of vanadate on glucose metabolism via the regulation of intracellular and intravesicular pH should be further pursued.

Acknowledgements

Y.S. is the incumbent of the C.H. Hollenberg Chair in Metabolic and Diabetes Research established by the Friends and Associates of Dr. C.H. Hollenberg of Toronto, Canada.

References

Amir, S. *et al.*: 1987, 'Vanadate ions. Central nervous system action on glucoregulation', *Brain Res.* **419**, 392-396.

Berneir, M. et al.: 1988, 'Effect of vanadate on the cellular accumulation of PP15, an apparent product of insulin receptor tyrosine kinase action', J. Biol. Chem. 263, 13626-13634.

Blondel, O. et al.: 1989, 'In vivo insulin resistance in streptozotocin-diabetic rats – evidence for reversal following oral vanadate treatment', Diabetologia 32, 185-190.

Canalis, E.: 1985, 'Effect of sodium vanadate on deoxyribonucleic acid and protein synthesis in cultured rat valvariae', Endocrinology 116, 855-862.

Cantley, L.C. et al.: 1978, 'Vanadate inhibits the red cell (Na$^+$,K$^+$)ATPase from the cytoplasmic side', Nature 272, 552-554.

Cassel, D. et al.: 1984, 'Vanadate stimulated Na$^+$,H$^+$ exchange activity in A431 cells', Biochem. Biophys. Res. Commun. 118, 675-681.

Clark, A.S. et al.: 1985, 'Selectivity of vanadate on glucose and protein metabolism in skeletal muscle', Biochem. J. 232, 173-179.

Clausen, T. et al.: 1981, 'The relationship between the transport of glucose and cations across cell membranes in isolated tissues. XI. The effect of vanadate on ^{45}Ca-efflux and sugar transport in adipose tissue and skeletal muscle', Biochim. Biophys. Acta 646, 261-267.

Czech, M.P.: 1980, 'Insulin action and the regulation of hexose transport', Diabetes 29, 399-409.

Czech, M.P.: 1985, 'The nature and regulation of the insulin receptor: Structure and function', Ann. Rev. Physiol. 47, 357-381.

Degani, H. et al.: 1981, 'Electron paramagnetic studies and insulin-like effects of vanadium in rat adipocytes', Biochemistry 20, 5795-5799.

Delfert, D.M. and McDonald, J.M.: 1985, 'Vanadyl and vanadate inhibit Ca^{2+} transport systems of the adipocyte plasma membrane and endoplasmic reticulum', Arch. Biochem. Biophys. 241, 665-672.

Dlouha, H. et al.: 1981, 'The effect of vanadate on the electrogenic Na$^+$/K$^+$ pump, intracellular Na$^+$ concentration and electrophysiological characteristic of mouse skeletal muscle fiber', Physiol. Bohemoslov. 30, 1-10.

Drust, D.S. and Creutz, C.E.: 1988, 'Aggregation of chromaffin granules by calpactin at micromolar levels of calcium', Nature 331, 88-91. Dubyak, G.R. and Kleinzeller, A.: 1980, 'The insulin-mimetic effects of vanadate as a Na$^+$,K$^+$)ATPase inhibitor', J. Biol. Chem. 255, 5306-5312.

Duckworth, W.C. et al.: 1988, 'Insulin like effects of vanadate in isolated rat adipocytes', Endocrinology 122, 2285-2289.

Fagin, J.A. et al.: 1987, 'Insulinotropic effects of vanadate', Diabetes 36, 1448-1452.

Gil, J. et al.: 1988, 'Insulin-like effects of vanadate on glucokinase activity and fructose 2.6 biphosphate levels in the liver of diabetic rats', J. Biol. Chem. 263, 1868-1871.

Gil, J. et al.: 1986, 'Effect of diabetes on fructose 2.2-P$_2$, glucose 1.6-P$_2$ and 6-phosphofructo 2-kinase in rat liver', Biochem. Biophys. Res. Commun. 136, 498-503.

Heyliger, C.E. et al.: 1985, 'Effects of vanadate on elevated blood glucose and depressed cardiac performance of diabetic rats', Science 227, 1474-1476.

Heyliger, C.E. et al.: 1985, 'Effect of vanadate on elevated blood glucose and depressed cardiac performance of diabetic rats', Science 227, 1474-1476.

Hirsova, D. and Koldovsky, O.: 1969, 'On the question of the absorption of insulin from the gastrointestinal tract during postnatal development', Physiolog. Bohemoslov. 18, 281-284.

Hori, C. and Oka, T.: 'Vanadate enhances the stimulatory action of insulin on DNA synthesis in cultured mouse mammary gland', Biochim. Biophys. Acta 610, 235-240.

Jackson et al.: 1988, 'Insulin-mimetic effects of vanadate in primary cultures of rat hepatocytes', Diabetes 37, 1234-1240.

James, D.E. et al.: 1988, 'Insulin regulatable tissues express a unique insulin-sensitive glucose transport protein', Nature 333, 183-185.

Kadota, S. et al.: 1987, 'Peroxide(s) of vanadium: A novel and potent insulin-mimetic agent which activates the insulin receptor kinase', Biochem. Biophys. Res. Commun. 147, 259-264.

Kadota, S. et al.: 1986, 'Vanadate stimulation of IGF-I binding to rat adipocytes', Biochem. Biophys. Res. Commun. 138, 174-178.

Kahn, C.R. and White, M.F.: 1988, 'The insulin receptor and the molecular mechanism of insulin action', J. Clin. Invest. 82, 1151-1156.

Kahn, C.R. and Shechter, Y.: 1990, 'Insulin, oral hypoglycemic agents and pharmacology of the endocrine pancreas', in *Goodman and Gilman Textbook of Pharmacology*, in press.

Karnieli, E. *et al.*: 1981, 'Insulin-stimulated translocation of glucose transport systems in the isolated rat adipose cell. Time course, reversal, insulin concentration dependency, and relationship to glucose transport activity', *J. Biol. Chem.* **256**, 4772-4782.

Kasuga, M.Y. *et al.*: 1978, 'Insulin binding and glucose metabolism in adipocytes of streptozotocin-diabetic rats', *Am. J. Physiol.* **235**, E175-E182.

Kasuga, M. *et al.*: 1982, 'Insulin stimulates the phosphorylation of the 95,000 dalton β-subunit', *Science* **215**, 185-187.

Kobayashi, M. and Olefsky, J.M.: 1979, 'Effect of streptozotocin induced diabetes on insulin binding, glucose transport and intracellular glucose metabolism in isolated rat adipocytes', *Diabetes* **28**, 87-95.

Kono, T. *et al.*: 1982, 'Evidence that translocation of the glucose transport activity is the major mechanism of insulin action on glucose transport in fat cells', *J. Biol. Chem.* **257**, 10942-10947.

Lindquist, R.N. *et al.*: 1973, 'Possible transition-state analogs for ribonuclease. The complexes of uridine with oxovanadium (IX) ion and vanadium (V) ion', *J. Am. Chem. Soc.* **95**, 8762-8768.

Lopez, V. *et al.*: 1976, 'Vanadium ion inhibition of alkaline phosphatase-catalyzed phosphate ester hydrolysis', *Arch. Biochem. Biophys.* **175**, 31-38.

Macara, I.G.: 1980, 'Vanadium – an element in search of a role', *Trends Biochem.* **5**, 92-94.

Meyerovitch, J. *et al.*: 1987, 'Oral administration of vanadate normalizes blood glucose levels in streptozotocin-treated rats, characterization and mode of action', *J. Biol. Chem.* **262**, 6658-6662.

Meyerovitch, J. *et al.*: 1989, 'Vanadate stimulated *in vivo* glucose uptake in brain tissue and arrests food consumption and weight gain', *Physiology and Behaviour*, in press.

Mondon, C.E. *et al.*: 1974, 'Insulin sensitivity of isolated perfused rat liver', *Diabetes* **24**, 25-229.

Moore, R.D.: 1983, 'Effects of insulin upon ion transport', *Biochim. Biophys. Acta* **73**, 1-49.

Nechay, B.R.: 1984, 'Mechanism of action of vanadium', *Ann. Rev. Pharmacol.* **24**, 501-524.

Pederson, R.A. *et al.*: 1989, 'Long term effects of vanadyl treatment on streptozotocin-induced diabetes in the rat', *Diabetes*, in press.

Pershadsingh, H.A. *et al.*: 1987, 'Chelatin of intracellular calcium prevents stimulation of glucose transport by insulin and insulinomimetic agents in the adipocyte. Evidence for a common mechanism', *Endocrinology* **121**, 1727-1732.

Post, R.L. *et al.*: 1979, 'Vanadium compounds in relation to inhibition of sodium and potassium adenosine triphosphatase', *Proceedings of the 2nd International Conference on the Properties and Functions of Na, K-ATPase. Academic*, New York, pp. 389-401.

Ramasarma, T. and Crane, F.L.: 1981, 'Does vanadium play a role in cellular regulation?' *Curr. Top. Coll. Regul.* **20**, 247-301.

Reddy, S.S.K. and Kahn, C.R.: 1988, 'Insulin resistance: A look at the role of insulin receptor kinase', *Diabetic Medicine* **5**, 621-629.

Reid, T.W. and Reid, W.A.: 1987, 'The labile nature of the insulin signal(s) for the stimulation of DNA synthesis in mouse lens epithelial and 3T3 cells', *J. Biol. Chem.* **262**, 229-233.

Shechter, Y. and Karlish, S.J.D.: 1980, 'Insulin-like stimulation of glucose oxidation in rat adipocytes by vanadyl (IV) ions', *Nature* **284**, 556-558.

Shechter, Y.: 1984, 'Trifluperazine inhibits insulin action on glucose metabolism in fat cells without affecting inhibition of lipolysis', *Proc. Natl. Acad. Sci. USA* **81**, 327-331.

Shechter, Y. and Henis, Y.: 1984, 'Cis-unsaturated fatty acids induce both lipogenesis and calcium binding in adipocytes', *Biochim. Biophys. Acta* **805**, 89-96.

Shechter, Y.: 1985, 'Studies on insulin receptors: Implication for insulin action', in *The Receptors*, Academic Press, Vol. II, pp. 221-224.

Shechter, Y. and Ron, A.: 1986a, 'Effect of depletion of phosphate and bicarbonate ions on insulin action in rat adipocytes', *J. Biol. Chem.* **261**, 14945-14950.

Shechter, Y. and Ron, A.: 1986b, 'Effect of depletion of bicarbonate or phosphate ions on insulin action in rat adipocytes. Further characterization of the receptor effector system', J. Biol. Chem. **261**, 14951-14945.

Shechter, Y. *et al.*: 1988, 'The use of post-binding agents in studying insulin action and its relation to experimental diabetes', *Biochem. Pharmacol.* **37**, 1891-1896.

Simons, T.J.B.: 1979, 'Vanadate – a new tool for biologists', *Nature*, **281**, 337-338.

Simpson, I.A. and Cushman, S.W.: 1986, 'Hormonal regulation of mammalian glucose transport', *Ann. Rev. Biochem.* **55**, 1059-1089.

Smith, J.B.: 1983, 'Vanadium ions stimulate DNA synthesis in Swiss mouse 3T3 and 3T6 cells', *Proc. Natl. Acad. Sci. USA* **80**, 6162-6166.

Stralfors, P.: 1988, 'Insulin stimulation of glucose uptake can be mediated by diacylglycerol in adipocytes', *Nature* **335**, 554-556.

Suzuki, K. and Kono, T.: 1980, 'Evidence that insulin causes translocation of glucose transport activity of the plasma membrane from an intracellular storage site', *Proc. Natl. Acad. Sci. USA* **77**, 2542-2545.

Swarup, G. *et al.*: 1982, 'Phosphotyrosyl-protein phosphatase of TCRC-2 cells', *J. Biol. Chem.* **257**, 7298-7301.

Szabo, A.J. and Szabo, O.: 1983, 'Insulin injected into CNS structures or into the carotid artery: Effect on carbohydrate homeostasis of the intact animal', *ADv. Metab. Disorders* **10**, 385-400.

Takayama, S. *et al.*: 1988, 'Phorbol ester induced serine phosphorylation of the insulin receptor decreases its tyrosine kinase activity', *J. Biol. Chem.* **263**, 3440-3447.

Tamura, S. *et al.*: 1984, 'A novel mechanism of the insulin-like effects of vanadate on glycogen synthase in rat adipocytes', *J. Biol. Chem.* **259**, 6650-6658.

Thomas, R.C.: 1989, 'Bicarbonate and pH response', *Nature* **337**, 601.

Weinhouse, S.: 1976, 'Regulation of glucokinase in liver', *Curr. Top. Cell Resul.* **11**, 1-50.

Werden, K. *et al.*: 1982, 'Stimulatory (insulin-mimetic) and inhibitory (ouabain-like) action of vanadate on potassium uptake and cellular sodium and potassium in heart cells in culture', *Biochim. Biophys. Acta* **23**, 79-93.

White, M.F. *et al.*: 1988, 'A cascade of tyrosine autophosphorylation in the β-subunit activates the phosphotransferase of the insulin receptor', *J. Biol. Chem.* **263**, 2969-2980.

VIII. Vanadate Sensitized Photocleavage of Proteins

I. R. GIBBONS and GABOR MOCZ

Pacific Biomedical Research Center, University of Hawaii, Honolulu HI 96822, U.S.A.

Over the past three years, it has become clear that vanadate, the vanadium(V) oxyanion, can serve as a photosensitizing chromophore that is capable of catalyzing the scission of peptide bonds at specific locations within a protein. This photolytic activity of vanadate is closely related to its ability to inhibit many of the enzymes involved in phosphate metabolism. In most cases, this inhibition occurs as a result of the tetrahedral VO_4^{3-} anion acting as a phosphate analogue that binds to a phosphate binding site of an enzyme with substantially greater affinity than phosphate itself and so acting as a dead-end kinetic block. Among the phosphohydrolases inhibited in this way are acid and alkaline phosphatases, RNAase, some protein phosphatases, the ion-transporting Ca^{2+}- and Na,K-ATPases in cell membranes, the dynein ATPases responsible for many forms of microtubule-based motility, and the myosin ATPase responsible for relative sliding of myosin and actin filaments in muscle (van Etten *et al.*, 1974; Lopez *et al.*, 1976; Lindquist *et al.*, 1973; Cantley *et al.*, 1977; Gibbons *et al.*, 1978; Kobayashi *et al.*, 1978; Goodno, 1979; Markus *et al.*, 1989). These enzymes differ substantially in their manner of interaction with phosphate, and the detailed mechanism of their inhibition by vanadate presumably differs correspondingly (Boyd and Kustin, 1984). Oligomeric forms of vanadate can also bind to specific sites on some proteins, presumably through a different mechanism than monovanadate, and decavanadate is a moderately potent inhibitor of a variety of phosphotransferases, including hexokinase, adenylate kinase and phosphofructokinase, that are not inhibited by monomeric vanadate (Boyd *et al.*, 1985).

The first description of site-directed photolysis mediated by vanadate was given in a study showing that irradiation at 254 nm of dynein ATPase from sperm flagella in the presence of low micromolar concentrations of monomeric vanadate and $MgATP^{2-}$ cleaved the α and β heavy chain (470 kDa) polypeptides of the dynein at a specific site, termed the V1 site, to give cleavage peptides of 250 kDa and 220 kDa with a conversion efficiency of 63% (Lee-Eiford *et al.*, 1986). However, irradiation of the protein at this wavelength resulted in the specific photocleavage being accompanied by a significant amount of nonspecific damage due to absorption by the aromatic amino acids. A major improvement in specificity was obtained by increasing the wavelength of irradiation to 365 nm, for at this wavelength the vanadate retains a significant absorbance of about 30 M^{-1} cm^{-1} (Boyd and Kustin, 1984) whereas the absorbance of the aromatic amino acids is negligible. Upon irradiation at 365 nm in the presence of monomeric vanadate and $MgATP^{2-}$, both the α and β heavy chain polypeptides of the dynein were cleaved at their V1 site with no apparent side effects, and the recovery of specific cleavage

N. Dennis Chasteen (ed.), Vanadium in Biological Systems, 143–152.
© 1990 *Kluwer Academic Publishers. Printed in the Netherlands.*

peptides was better than 90% (Gibbons et al., 1987). Further exploration of the vanadate-mediated photolytic reaction of dynein showed that irradiation at 365 nm in the presence of oligomeric vanadate and Mn^{2+}, and in the absence of ATP, cleaved the α and β heavy chains at a different site, termed the V2 site, that was located 70-100 kDa toward the amino terminus from the V1 site (Tang and Gibbons 1987).

More recent studies have shown that sensitivity to vanadate-mediated photolysis appears to be a general property of dyneins from many, possibly all, sources. The properties of photocleavage at the V1 site appear to be particularly highly conserved, occurring in all seven isoforms of dynein heavy chain from sea urchin sperm (Gibbons B. H. and Gibbons, 1987), in flagellar dyneins from *Chlamydomonas* and in ciliary dyneins from *Tetrahymena* (King and Witman, 1987, 1988; Gibbons I. R. and Gibbons, 1987; Marchese-Ragona et al., 1989), as well as in cytoplasmic dyneins from mammalian brain, nematodes, *Drosophila*, and sea urchin eggs (Vallee et al., 1988; Lye et al., 1987; Porter et al., 1988). The detailed properties of the V1 cleavage reaction appear to differ only slightly among species, with the most notable variation observed so far being that photolysis at the V1 site of the β heavy chain of dynein from higher animals occurs only in the presence of ATP or ADP, whereas that in the β chain from *Chlamydomonas* and *Tetrahymena* occurs also in the absence of nucleotide. The properties of the V2 cleavage reaction appear to show more variation between species and between different isoforms of dynein, indicating that this aspect of the ATP- binding site has been less tightly conserved during evolution than that of the V1 site. In dynein from sea urchin flagella, the V2 cleavage of the α chain occurs about four times faster than that of the β chain (Tang and Gibbons, 1987). Moreover, in dyneins from *Chlamydomonas*, V2 cleavage of the α and γ heavy chains occurs at 2-3 separate sites up to 40 kDa apart, although the cleavage of the β chain occurs at a unique site as in sea urchin (King and Witman, 1987, 1988).

Although the heavy chains of dynein ATPase appear to be uniquely sensitive to vanadate-mediated photolysis, a variety of other proteins can be photocleaved under appropriate conditions. Irradiation in the presence of oligomeric vanadate produces photocleavage at three distinct sites on the heavy chains of myosin from rabbit skeletal muscle, suggesting that these cleavage sites may correspond to phosphate-binding sites on the myosin heavy chain (Mocz, 1989). Under other conditions, irradiation in the presence of monomeric vanadate can cleave myosin subfragment 1 at a single site in a two-step reaction in which a serine at the active site is first photooxidized to serine aldehyde, followed by a second vanadate-promoted photoreaction that cleaves the peptide backbone (Cremo et al., 1988; Grammer et al., 1988). Irradiation of D-ribulose-1,5-bisphosphate carboxylase/oxygenase from spinach in the presence of vanadate also results in photomodification of a serine at the active site, with subsequent photocleavage in the large subunit of the enzyme (Mogel and McFadden, 1989). Vanadate-mediated photolysis has also been observed in pyruvate kinase from rabbit and chicken muscle (Mocz, G., unpublished data).

Vanadate is not unique in its ability to catalyze photolytic scission of polypeptide chains, and the ATP- and ADP-complexes of at least two other transition metals, iron(III) and rhodium(III), behave similarly (Mocz and Gibbons, 1990). In these instances, it is likely that the site-directed photolysis involves the transition metal cation substituting for the usual Mg^{2+} cation in the enzyme-nucleotide complex. Some of the sites of cleavage appear to be the same as those obtained with vanadate, but others are different.

As an example of the properties of vanadate-mediated photolysis, we now describe in greater detail its application to the outer arm dynein from sea urchin sperm flagella.

Properties of Vanadate-Mediated Cleavage of Dynein from Sea Urchin Sperm Flagella

CLEAVAGE AT THE V1 SITE IN THE PRESENCE OF MONOMERIC VANADATE

The standard medium used for photolysis of dynein heavy chains contains 0.45 M sodium acetate, 2.5 mM magnesium acetate, 0.5 mM EDTA, 50 μM vanadate, 100 μM ATP, and 10 mM HEPES/NaOH buffer, pH 7.5. Irradiation is performed with a Model EN-28 lamp (Spectronics Corp., Westbury, NY) having a spectrum consisting of a single emission line at 365 nm, superimposed upon a background continuous spectrum extending from 320 to 400 nm with a broad maximum at 355 nm of intensity about 30% less than that of the 365 nm line. The total intensity of irradiation is ~ 2 mWatt.cm^{-2} at the position of the sample. Under these conditions, cleavage of the α and β heavy chains occurs specifically at the V1 site with a yield of more than 90% (Figure 1). The rate of cleavage has a biphasic dependence upon time, with $\sim 80\%$ of the dynein being cleaved with a $t_{1/2}$ of 7 min, and the remainder with a $t_{1/2}$ of ~ 90 min. The ATPase activity of the dynein is lost in parallel with the cleavage of the heavy chains (Figure 2), strongly suggesting that its loss is a direct consequence of the cleavage reaction. The cleaved dynein retains normal affinity for binding to dynein-depleted sperm flagella (Gibbons and Gibbons, 1987) and is also able to bind ATP, but it has no hydrolytic activity. The rate of cleavage at the V1 site shows a hyperbolic dependence upon vanadate concentration, with half-maximal rate occurring at a concentration of ~ 4.5 μM (Figure 3), consistent with the chromophore being the inhibitory vanadate bound at the γ-P_i locus of the hydrolytic ATP binding site in the inhibited dynein.ADP.vanadate complex. The action of vanadate in photocleavage is catalytic, and substoichiometric concentrations of vanadate (5-10% of that of the dynein heavy chains) also promote the photolysis.

Cleavage is usually performed in the presence of Mg^{2+}, and little or no cleavage occurs in the absence of any added divalent cations. The Mg^{2+} can be replaced by either Ca^{2+}, or Zn^{2+} with nearly equal effectiveness. Substitution of Mg^{2+} by any of the transition metal cations Mn^{2+}, Fe^{2+}, or Co^{2+} suppresses the photocleavage, presumably by quenching the excited chromophore prior to scission of the heavy chains. However, the cleavage is not affected significantly

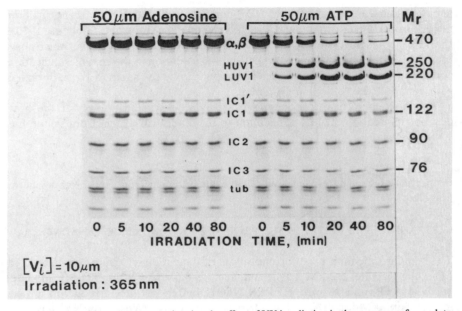

Fig. 1. Gel electrophoretic patterns showing the effect of UV irradiation in the presence of vanadate and MgATP on the polypeptides of dynein from sea urchin sperm flagella as a function of time. The samples were irradiated in medium containing either 50 μM adenosine (left six lanes) or 50 μM ATP (right six lanes). The irradiation medium also contained 10 μM vanadate, 0.45 M Na acetate, 2.5 mM Mg acetate, 0.5 mM EDTA, 7 mM 2-mercaptoethanol, 10 mM HEPES/NaOH buffer, pH 7.5. Molecular weights (in thousands) of the dynein subunits and of the newly formed cleavage peptides (HUV1 and LUV1) are given in the right margin. The percentage of cleavage at each irradiation time was determined by extraction of dye from the appropriate electrophoretic bands; the results are shown together with the corresponding ATPase activities in Fig. 2 (Gibbons et al., 1987).

by addition of general free radical trapping agents such as 0.1 M cystamine, or 2-amino-ethylcarbamidothioic acid dihydrobromide. This indicates that the quenching of cleavage by Mn^{2+}, Fe^{2+}, and Co^{2+} is a specific effect, and that the cleavage reaction is not due to a kinetically accessible free radical.

The effectiveness of different nucleotides in supporting V1 cleavage roughly parallels their ability to act as substrates for dynein ATPase. In the presence of 20 μM vanadate, CTP and UTP support cleavage at about half the rate of ATP, whereas GTP and ITP support cleavage only if the vanadate concentration is raised to about 200 μM.

The rate of cleavage at the V1 site has only a low dependence upon temperature. The initial rate of cleavage when the irradiation is performed in a frozen medium containing 20% sucrose at -78°C is only about 4-fold slower than that at the usual temperature of 8°C (Figure 4). It is only when the temperature is lowered to -196°C that the rate of cleavage is substantially decreased. This low dependence on temperature is typical of a free radical reaction.

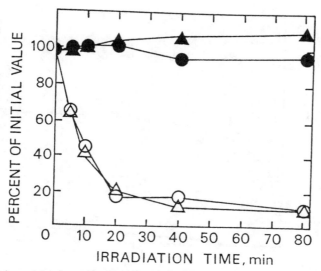

Fig. 2. Comparison of the cleavage of dynein heavy chains at their V1 sites with the loss of ATPase activity for the samples illustrated in Fig. 1. Percentage of remaining intact α and β heavy chains: ○, sample irradiated in 10 μM vanadate and 50 μM ATP; ●, sample irradiated in 10 μM vanadate and 50 μM adenosine. Percentage remaining of the initial ATPase activity: △, sample irradiated in 10 μM vanadate and 50 μM ATP; ▲, sample irradiated in 10 μM vanadate and 50 μM adenosine (Gibbons *et al.*, 1989).

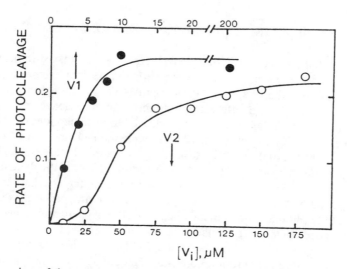

Fig. 3. Comparison of cleavage rates at the V1 and the V2 sites as functions of vanadate concentration. The upper line shows the cleavage rate at the V1 site, obtained by irradiating dynein as described in Fig. 1, with the concentrations of vanadate indicated on the upper scale. The lower line shows the rate of cleavage at the V2 site, obtained by irradiating dynein under V2 conditions as described in Fig. 4, with the concentrations of vanadate indicated on the lower scale (Gibbons *et al.*, 1989).

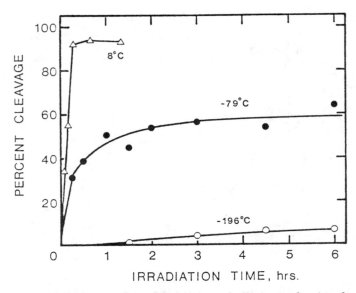

Fig. 4. The rate of photocleavage of dynein heavy chains at the V1 site as a function of temperature. Conditions were: irradiated at 8°C in standard medium as described in Fig. 1; irradiated immersed in acetone-dry ice at -78°C after first being frozen with liquid nitrogen in standard medium containing 30% sucrose; irradiated immersed in liquid nitrogen at -196°C in standard medium containing 30% sucrose (Previously unpublished data of Dr. B. H. Gibbons).

CLEAVAGE AT THE V2 SITE IN THE PRESENCE OF OLIGOVANADATE

When the irradiation medium is changed by increasing the vanadate concentration to 100 μM, substituting manganese acetate for magnesium acetate in order to suppress cleavage at the V1 site, and omitting the ATP, irradiation under the same physical conditions yields a different pattern resulting from cleavage at the V2 site on the dynein heavy chains. The α chain is cleaved to form peptides of ~ 280 kDa and ~ 190 kDa with a $t_{1/2}$ of about 12 min, while the β chain is cleaved to form peptides of 275 kDa and 195 kDa with a $t_{1/2}$ of about 50 min. Unlike cleavage at the V1 site, cleavage at the V2 site does not result in loss of the dynein ATPase activity (Figure 5). Inclusion of 5 mM glutathione in the irradiation medium provides partial protection against changes in ATPase properties caused by side reactions of the photoactivated metal. The rate of heavy chain cleavage shows a sigmoidal dependence upon vanadate concentration, with half-maximal rate occurring at 58 μM and a sigmoidicity of 2.7 (Figure 3). The sigmoidicity of 2.7 strongly suggests that the chromophore is a vanadate oligomer, presumably tri or tetravanadate. At vanadate concentrations above 100 μM, the specificity of cleavage for the V2 site diminishes, so that some cleavage also occurs at V1 and other sites (Gibbons et al., 1989). Addition of 10 μM ATP, or ADP, or of 100 μM CTP or UTP to the irradiation medium inhibits cleavage at the V2 site, and results in a slow cleavage at the V1 site.

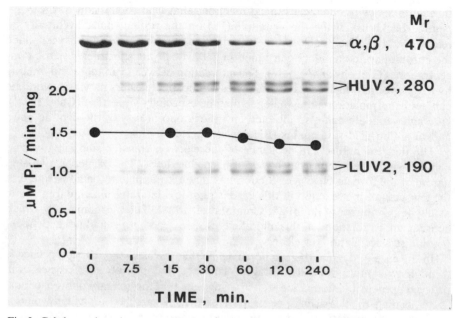

Fig. 5. Gel electrophoresis patterns showing the time course of cleavage of the dynein heavy chains at their V2 sites. Dynein was irradiated in medium containing 100 μM vanadate, 1 mM manganese acetate, 0.45 M sodium acetate, 0.1 mM EDTA, 5 mM glutathione, 1 mM NaSO$_4$, 10 mM HEPES/NaOH buffer, pH 7.4. At each of the times indicated samples were removed for electrophoretic analysis and assay of ATPase activity. α and β indicates the intact α and β heavy chains, which comigrate in this gel system; HUV2 and LUV2 indicate the V2 cleavage peptides. Specific ATPase activity of the irradiated samples is indicated by the line with $\oplus$. Molecular weights (in thousands) are indicated on right (Tang and Gibbons, 1987).

A variety of free radical trapping agents, including sodium azide, thiourea, Tris, cystamine (all at 1-10 mM) have no inhibitory effect on the cleavage, indicating that the cleavage at the V2 site, like that at the V1 site, is due to localized electron transfer without the involvement of kinetically accessible free radicals.

Mechanism of Photocleavage

Although the detailed basis for the photolytic activity of vanadate is unknown, it is reasonable to postulate that the initial step involves photoexcitation of the electronic structure of the metal ion while it is complexed with the divalent cation and with one or more amino acid side chains on the protein, as well as with the nucleotide when it is present. Subsequent acceptance of an electron from an adjacent group on the protein could be facilitated by the strong tendency of reduced vanadium(IV) to be stabilized by complexation with other carboxyl and amino residues on the protein (Chasteen, 1983; Chasteen et al., 1986). The oxidized radical group thus generated on the protein presumably then proceeds through a series of chemical radical reactions that lead to scission of the polypeptide backbone, while the vanadium(IV) on the protein must eventually be released and oxidized back to its + 5 oxidation state by molecular oxygen since the chromophoric vanadate can function catalytically at substoichiometric concentrations (Gibbons et al., 1987). The interaction between vanadate and organic hydroxyl groups which leads to analogues of phosphate esters in which vanadate can act as a transition-state analogue may be especially important (Chapter IV), although vanadate can also interact with the amino moiety of the peptide bond and of amino acid side chains (Rehder et al., 1988).

The detailed nature and sequence of the electron transfer steps involved in vanadate-mediated photolysis have not been elucidated. The particular two-step nature of the myosin cleavage reaction has made it possible to demonstrate that the first step of the reaction in this case is photooxidation of a serine to a serine aldehyde (Grammer et al., 1988; Cremo at al, 1988). The mechanism by which subsequent irradiation of this modified protein leads to vanadate dependent peptide scission is not yet clear.

In the case of dynein, there is no evidence that the cleavage reaction proceeds through two photo-dependent stages. The fact that photolysis of dynein will proceed upon irradiation in ice at -78°C suggests that the primary photochemical event leading to cleavage is generation of a localized free radical by photo-activation of vanadate bound at the catalytic site on the heavy chains (Gibbons et al., 1988). As mentioned above, the lack of quenching by free radical scavengers in both the V1 and the V2 cleavage reactions also indicates that the cleavage process involves electron transfer in the locality of the primary chromophore. The quenching of cleavage at the V1 site by Mn^{2+} presumably occurs through a redox reaction of the Mn^{2+} bound at the catalytic site that transfers the energy of the excited vanadate chromophore to the solvent.

On general principles it seems probable that O_2 plays a major role in the mechanism by which electron excitation leads to polypeptide scission. When irradiation of dynein under V1 conditions is performed in the absence of molecular oxygen in an argon filled spectrophotometric cell, little or no cleavage of the heavy chains occurs (Ungacta, F. and Mocz, G., unpublished data). One possibility is that molecular oxygen combines with the photoactivated dynein.ADP.vanadate complex that then undergoes a free radical reaction in an inner electron sphere. Alternatively, the mechanism may be involve photo-oxidation of a serine residue as found in myosin (Cremo et al., 1988). In either case, however, it will be necessary to account for the lack of new α-amino-terminal groups in the cleavage reaction in both myosin and dynein. Scission of the polypeptide backbone could result from vanadate catalyzed oxidative decarboxylation of an acidic side chain (Dutta et al., 1987) leading to cleavage of a peptide bond, or from production of α-carbon radicals that decompose in the presence of oxygen to cleave the polypeptide chain at the α carbon (with production of a carbonyl and an amide) rather than at the peptide bond (Garrison et al., 1962).

Applications of Photocleavage

Because the photocleavage of polypeptides can be performed under conditions that are close to physiological, it offers a useful new strategy for probing the structure of the substrate binding site of certain enzymes and for obtaining information about changes in their conformation. Present applications of vanadate-mediated photocleavage include: (a) identification of high molecular weight polypeptides as subunits of a dynein ATPase by cleavage at their V1 site (Gibbons *et al.*, 1989); (b) linear mapping of dynein heavy chains through combination of photolysis with limited tryptic digestion and monoclonal antibody markers (Mocz *et al.*, 1988, King and Witman 1988); (c) indication of a conformational change in the vicinity of the ATP binding site (Marchese-Ragona *et al.*, 1989); (d) identification of amino acid residues at a phosphate binding site (Cremo *et al.*, 1988). In the future, the use of substrate analogues containing a photosensitizing transition metal at an appropriate location may significantly expand the range of enzymes whose conformational changes can be examined through sensitivity to other photo-modifications of amino acid side chains in reactions that either may or may not proceed to peptide cleavage.

Acknowledgements

This work has been supported NIH Grants GM30401 (to IRG) and HD06565 (to Dr. Barbara H. Gibbons).

References

Boyd, D. W. and Kustin, K. (1984) *in* Advances in Inorganic Biochemistry (Eichorn, G. L. and Marzilli, L. G., eds) Vol. 6, pp 311-365, Elsevier Scientific Publishing Co., New York

Boyd, D. W., Kustin, K. and Miwa, M. (1985) *Biochim. Biophys. Acta* **827**, 472- 475.

Cantley, L. C., Josephson, L., Warner, R., Yanagisawa, M., Lechene. C. and Guidotti, G. (1977) *J. Biol. Chem.* **252**, 7421-7433.

Chasteen, N. D. (1983) *Structure and Bonding* **53**, 105-138.

Chasteen, N. D., Grady, J. K. and Holloway, C. E. (1986) *Inorg. Chem.* **25**, 2754-2760.

Cremo, C. R., Grammer, J. C. and Yount, R. G. (1988) *Biochemistry* **27**, 8415-8420.

Dutta, H., Hazra, B., Banerjee, A. and Banerjee, F. (1987) *J. Indian Chem. Soc.* **64**, 706-707.

Garrison, W. M., Jayko, M. E. and Bennett, W. (1962) *Radiation Research* **16**, 483-502.

Gibbons, B. H. and Gibbons, I. R. (1987) *J. Biol. Chem.* **262**, 8354-8359.

Gibbons, B. H., Tang, W.-J. Y. and Gibbons, I. R. (1988) *Cell. Motil. Cytoskel.* **11**, 188-189.

Gibbons, I. R., Cosson, M. P., Evans, J. A., Gibbons, B. H., Houck, B., Martinson, K. H., Sale, W. S. and Tang W.-J. Y. (1978) *Proc. Natl. Acad. Sci. U.S.A.* **75**, 2220-2224.

Gibbons, I. R. and Gibbons, B. H. (1987) *in* Perspectives of Biological Transduction. (Mukahata, Y., Morales, M. F. and Fleischer, S. eds) Academic Press, Tokyo. pp 107-116.

Gibbons, I. R., Lee-Eiford, A., Mocz, G., Phillipson, C. A., Tang, W.-J., Y. and Gibbons, B. H. (1987) *J. Biol. Chem.* **262**, 2780-2786.

Gibbons, I. R., Tang, W.-J. Y. and Gibbons, B. H. (1989) *in* Cell Movement, Vol. 1: The Dynein ATPases, (Warner, F. D., Satir, P. and Gibbons, I. R., eds) Alan R. Liss, Inc., New York, pp 77-88.

Goodno, C. C. (1979) *Proc. Nat. Acad. Sci. U.S.A.* **76**, 2620-2624.

Grammer, J. C., Cremo, C. R. and Yount, R. G. (1988) *Biochemistry* **27**, 8408-8415.

King, S. M. and Witman, G. B. (1987) J. Biol. Chem. **262**, 17596-17604.

King, S. M. and Witman, G. B. (1988) *J. Cell Biol.* **107**, 1799-1808.

Kobayashi, T., Martensen, T., Nath, J. and Flavin. M. (1978) *Biochem. Biophys. Res. Comm.* **81**, 1313-1318.

Lee-Eiford, A., Ow, R. A. and Gibbons, I. R. (1986) *J. Biol. Chem.* **261**, 2337- 2342.

Lindquist, R. N., Lynn, J. L. and Lienhard, G. E. (1973) *J. Am. Chem. Soc.* **95**, 8762-8768.

Lopez, V., Stevens, T. and Lindquist, R. N. (1976) *Arch. Biochem. Biophys.* **175**, 31-38.

Lye, R. J., Porter, M. E., Scholey, J. M. and McIntosh, J. R. (1987) *Cell* **51**, 309-318.

Marchese-Ragona, S. P., Facemeyer, K. C. and Johnson, K. A. (1989) *J. Cell Biol.* **109**, 157a.

Markus, S., Priel, Z. and Chipman, D. M. (1989) *Biochemistry* **28**, 793-799.

Mocz, G. (1989) *Eur. J. Biochem.* **179**, 373-378.

Mocz, G. and Gibbons, I. R. (1990) *J. Biol. Chem* (in press).

Mocz, G., Tang, W.-J. Y. and Gibbons, I. R. (1988) *J. Cell. Biol.* **106**, 1607-1614.

Mogel, S. N. and McFadden, B. A. (1989) *Biochemistry* **28**, 5428-5431.

Porter, M. E., Grissom, P. M., Scholey, J. M., Salmon, E. D. and McIntosh, J. R. (1988) *J. Biol. Chem.* **263**, 6759-6771.

Rehder, D., Weidemann, C., Duch, A. and Priebsch, W. (1988) *Inorg. Chem.* **27**, 584-587.

Tang, W.-J. Y. and Gibbons, I. R. (1987) *J. Biol. Chem.* **262**, 17728-17734.

Vallee, R. B., Wall, J. S., Paschal, B. M. and Shpetner, H. S. (1988) *Nature* **332**, 561-563.

van Etten, R. L., Waymock, P. P. and Rekhop, D. M. (1974) *J. Am. Chem. Soc.* **96**, 6782-6785.

IX. Vanadium in Ascidians

HITOSHI MICHIBATA
Biological Institute, Faculty of Science, Toyama University, Gofuku 3190, Toyama 930, Japan

and

HIROMU SAKURAI
Faculty of Pharmaceutical Sciences, University of Tokushima, Sho-machi 1, Tokushima 770, Japan

Introduction

About 80 years ago Henze (1911) found unexpectedly that ascidian blood cells contained a high level of vanadium. The initial interest in this phenomenon was not only the presence of vanadium but also the possibility that other metals might be concentrated in ascidians and other organisms. Many analytical chemists subsequently analyzed the vanadium content of various organisms (Cantacuzene and Tchekirian, 1932; Vinogradov, 1934; Kobayashi, 1935; Webb, 1939; Noddack and Noddack, 1939; Bertrand, 1950; Lybing, 1953; Boeri and Ehrenberg, 1954; Webb, 1956; Levine, 1961; Bielig *et al.*, 1961; Kalk, 1963a; 1963b; 1963c; Ciereszko *et al.*, 1963; Bielig *et al.*, 1966; Rummel *et al.*, 1966; Carlisle, 1968; Swinehart *et al.*, 1974; Danskin, 1978; Botte and Scippa, 1979; Michibata, 1984). Consequently, it became clear that the ascidians are the only organisms in animal kingdom known to accumulate high levels of vanadium and that all species among ascidians do not always contain large amounts of the metal.

Webb (1939) first classified ascidian species into three suborders, based on their vanadium content. That is to say, several species in the suborder Aplousobranchia have, a high vanadium content, and a significant amount of vanadium is found in representatives of the suborder Phlebobranchia, whereas species of the suborder Stolidobranchia contain relatively smaller amounts or little vanadium. Webb suggested that the presence of vanadium was a primitive characteristic which had been lost in more specialized families. Endean (1955a; 1960) also assumed that since the ascidians are an intermediate group between the invertebrates and vertebrates, these animals are the key to resolving how transition metals were originally selected by living organisms. Swinehart *et al.* (1974) suggested that the ascidian species might indeed represent animals that were in transition between the vanadium and iron "users". These earlier proposals have brought particular interest from a wide range of disciplines, including analytical chemistry, coordination chemistry, physiology and phylogeny.

Besides vanadium, niobium (Nb) (Carlisle, 1958; Kokubu and Hidaka, 1965), titanium (Ti) (Noddack and Noddack, 1939; Levine, 1961), chromium (Cr) (Levine, 1961; Botte *et al.*, 1979a) and tantalum (Ta) (Kokubu and Hidaka, 1965) have been reported to be present in ascidian tissues. However, the reproducibility of most of the data has been poor and the amounts of these metals detected are

N. Dennis Chasteen (ed.), Vanadium in Biological Systems, 153–171.
© 1990 *Kluwer Academic Publishers. Printed in the Netherlands.*

in very low concentrations. Recently, Roman *et al.* (1988), however, found higher levels of titanium than previously (1,552.5 mg/kg dry weight) in the blood cells of *Ascidia dispar*.

2. Recent Analysis of Vanadium in the Blood Cells

Previous data on the vanadium content of ascidian blood cells were reviewed by several investigators (Goodbody, 1974; Swinehart *et al.*, 1974; Biggs and Swinehart, 1976; Kustin *et al.*, 1983). Unfortunately the data reported previously were obtained using different materials, pretreatments and different analytical methods including colorimetry, emission spectrometry, and atomic absorption spectrometry which differ widely in sensitivity and precision for vanadium. Furthermore, the data were expressed in a per dry weight, per wet weight, per ash weight or per inorganic dry weight basis making comparisons impossible.

At the commencement of our study, we therefore re-examined the vanadium content of several tissues of ascidians employing the same pretreatment and using neutron activation analysis which is the most sensitive method for vanadium (Michibata *et al.*, 1986a) (Figure 1). Data on transition metals, vanadium, iron and manganese, obtained from several tissues of 15 species belonging two suborders, the Phlebobranchia and the Stolidobranchia. The data are in substantial agreement with previous reports of high vanadium levels in the family Ascidiidae, the highest amount being present in blood cells.

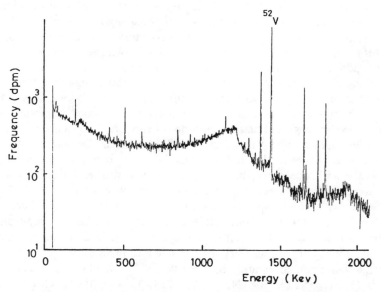

Fig. 1. γ-ray spectrometry of the blood cells of *Ascidia ahodori* after irradiation with thermal neutrons at the Institute for Atomic Energy, Rikkyo University. Clear photo-peak of ^{52}V, produced in the irradiated sample, can be seen. Neutron activation analysis is the most sensitive method for determination of vanadium (Michibata, 1989).

In contrast with the previous studies, however, we found significant levels of vanadium in all tissues of all species, even those belonging to the suborder Stolidobranchia. Differences in iron and manganese contents between the two suborders were not observed. In other words, relative concentrations of vanadium and iron in different ascidian subfamilies do not reflect phylogeny. Only the vanadium content differs markedly from each family or suborder. When the molar concentration of vanadium was calculated for tissues having 90% or greater moisture content, A. ahodori and A. sydneiensis samea belonging to Phlebobranchia were found to have 40 mM and 10 mM respectively vanadium in their blood cells. Recent data have revealed 150 mM vanadium in the blood cells of A. gemmata which also belongs to Phlebobranchia (Michibata et al., in preparation).

3. So-Called Vanadocyte, Vanadium-Containing Blood Cell

Ascidian blood cells are classified into six to nine different types based on their morphology (Wright, 1981). Among these, the morula cell has been thought to be involved in the accumulation of vanadium ions and has been called a vanadocyte because the green color of the cell resembles that of a vanadium complex, and dense granules observed after fixation with osmium tetroxide were (incorrectly) assumed to be deposits of vanadium (Webb, 1939; Boeri and Ehrenberg, 1954; Endean 1960; Kalk, 1963a; 1963b; 1963c).

Several investigators, mainly de Vincentiis' group, using X-ray microanalysis have claimed recently that the morula cell contains little or no vanadium ion (Botte et al., 1979b; Scippa et al., 1982; 1985; Rowley, 1982). They found more vanadium associated with the vacuolar membranes of granular amoebocyte, signet ring cells and compartment cell than in the vacuoles of the morula cells of Phallusia mammillata and Ciona intestinalis. Ultrastructural observations, however, can only provide limited data on the identification of the true vanadocyte because the quantitative analysis is done on two dimensions and there is the possibility of artifacts resulting from the preparation methods.

Therefore, we have attempted to identify the vanadocyte among several kinds of blood cells using a combination technique consisting of Ficoll density gradient centrifugation for purification of specific types of blood cells and neutron activation analysis for vanadium determination (Michibata et al., 1987). As shown in Figure 2, the distribution pattern of vanadium in four cell layers after the density gradient centrifugation is similar to that of the signet ring cell but is different from that of the other cell types. Thus, we have succeeded in showing that the vanadocyte is a signet ring cell, a vacuolated cell type, in A. ahodri and A. sydneiensis samea (Figure 3).

4. Differentiation of Blood Cells

The problems of which type of blood cell is the stem cell and how the stem cell differentiates to peripheral cells remain unsolved. It seems very important to know

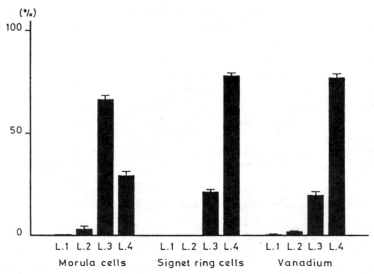

Fig. 2. Comparison of the patterns of distribution of morula cells and signet ring cells with that of vanadium ions after density gradient centrifugation. The pattern of distribution of vanadium is similar to that of signet ring cells but is different from the morula cells, proving clearly that the vanadocyte is signet ring cell. L.1: layer 1, L.2: layer 2, L.3: layer 3, L.4: layer 4 (Michibata *et al.*, 1987).

the lineage of the vanadocyte from the stem cell and when the accumulation of vanadium commences during the differentiation process of the blood cell. Besides many cytological and morphological studies on relationships between different blood cell types, Freeman (1964) and Ermak (1975) tried to clarify the differentiation process of ascidian blood cell by means of labeling experiment with tritiated thymidine and then suggested that the lymphocyte was the stem cell but vacuolated cells did not proliferate. Further details of the differentiation process is required. Through appropriate experimental studies using monoclonal antibodies and/or

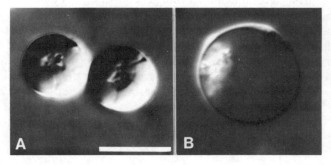

Fig. 3. Differential interference contrast photomicrographs of the morula cell (A), which has been thought to be involved in the accumulation of vanadium, and the signet ring cell (B), which has been newly identified as the vanadocyte, in *Ascidia ahodori*. Scale bar indicates 10 μm.

labeling compounds specific to some cytoplasmic substances available for ascidian blood (Nishide *et al.*, 1989).

Scippa *et al.* (1988) have reported that the vacuolated and granular amoebocytes, signet ring cells and a variety of compartment cells of *Phallusia mammillata* can be considered vanadocytes based on the data by X-ray microanalysis. There is also the opinion that the term vanadocyte encompasses all cells which assimilate vanadium (Smith, 1989). However, we propose that the term vanadocyte be used for the blood cell which is able to accumulate and/or contain a high amount of vanadium actually. Study of the differentiation process of blood cells will give a correct answer in time.

5. Fluorescence Derived from the Blood Cells

de Vincentiis and his co-workers (de Vincentiis, 1962; de Vincentiis and Rudiger, 1967) first noted the emission of fluorescence derived from ascidian blood cells and eggs. After that, Kustin's group (Macara *et al.*, 1979a; 1979b) guided by fluorescence measurements reported the successful isolation of a reducing agent from the blood cells of *A. nigra* and *C. intestinalis*. This substance, named tunichrome, has been reported to emit a specific autonomous fluorescence at 532 nm and 581 nm upon excitation with blue-violet light and to be able to reduced vanadate(V) ion to vanadyl(IV) ion and Fe(III) to Fe(II) in vitro. Thereafter, Nakanishi's group isolated a tunichrome B-1 from the blood cells of *A. nigra* under anaerobic conditions and verified its chemical structure (Bruening *et al.*, 1985). It was not until that time that they knew that the true vanadocyte was the signet ring cell, not the morula cell. Robinson *et al.* (1984) believed that the concentration of vanadium could be estimated from the intensity of fluorescence of the tunichrome in each type of blood cell because the tunichrome seemed to be present with vanadium at approximately equimolar concentrations.

In recent experiments, we found that the morula cell emitted autonomous fluorescence with longer wavelength than 515 nm after excitation with blue-violet light. However, no fluorescence due to the tunichrome was detected from the signet ring cell in *A. ahodori* as well as in *A. sydneiensis samea* (Michibata *et al.*, 1988; 1989a) (Figure 4).

Up to the present, there is no report of finding vanadium associated with tunichrome. Additionally, if tunichrome is involved in the accumulation and reduction of vanadium ion in ascidian blood cells, then it should coexist with vanadium in the signet ring cell.

Besides from blood cells, Deno (1987) has recently reconfirmed that ascidian eggs also emit fluorescence from the myoplasmic region of the cytoplasm. It will be very interesting to determine the substance(s) from which such autonomous fluorescence is derived.

158 H. MICHIBATA AND H. SAKURAI

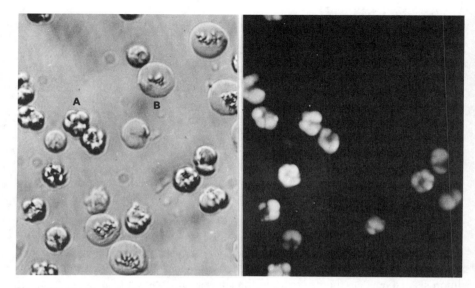

Fig. 4. Blood cells of *Ascidia ahodori* under a light and a fluorescence microscope. The morula cell (A) emits fluorescence due to the tunichrome upon excitation with blue-violet light but signet ring cell (B) does not fluoresce (Michibata *et al.*, 1988a; 1989).

6. Vanadium Complex

The chemical form of vanadium complex present in ascidian blood cells has long been a subject of discussion. Following Henze's finding (Henze, 1911) of vanadium in ascidian blood cells, it was believed that the vanadium was present in the form of a nitrogenous compound including sulfuric acid, well known as haemovanadin (Califano and Boeri, 1950; Webb, 1956; Bielig *et al.*, 1966). Kustin's group has claimed recently that the haemovanadin was an artificial compound produced by air oxidation. Since the vanadium ion dissolved in sea water is the vanadate(V) anion whereas that contained in ascidian blood cells is reduced to vanadyl cations, VO^{2+} and/or V^{3+}, it can be assumed that some reducing agents for reducing vanadate ion to vanadyl ion must exist in the blood cells (Macara, 1980).

We thought that vanadium must be present as a complex in ascidian blood cells and therefore attempted to extract and purify a vanadium-binding substance from ascidian blood cells although previous attempts had resulted in failure (Gilbert *et al.*, 1977; Macara *et al.*, 1979a; 1979b). Through a combination of techniques including chromatography, neutron activation analysis and ESR spectrometry, we succeeded in obtaining the vanadium-binding substance from the blood cells of *A. sydneiensis samea* under low pH conditions (Michibata *et al.*, 1986b) (Figure 5). This substance, named vanadobin, is colorless, can maintain the vanadium ion in the vanadyl form (VO(IV)) and has a specific affinity for vanadium ion.

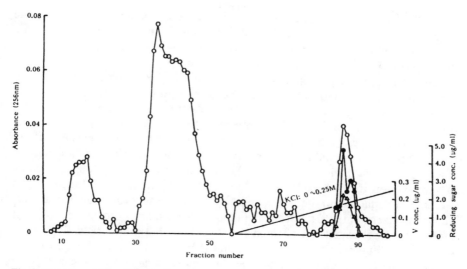

Fig. 5. Fraction profiles of vanadium-binding substance, named vanadobin. The blood cells of *Ascidia sydneiensis samea* were homogenized in 10 mM glycine-HCl buffer solution at pH 2.3. The homogenated was loaded onto the column of Sephadex G-25. The fractions containing vanadium were pooled and loaded again onto the column of SE-cellulose. The vanadobin was obtained with an elution of KCl solution. $\bigcirc$: Absorbance at 256 nm, $\triangle$: Vanadium concentration (μg/ml), $\bullet$: Reducing sugar concentration (μg/ml) (Michibata *et al.*, 1986b).

The most recent experiments have revealed that the vanadobin can be extracted from the blood cells of several vanadium containing species, *A. ahodori*, *A. zara*, *A. gemmata*, *Ciona intestinalis* and *Corella japonica* (Michibata *et al.*, in preparation). From cell fractionation studies, it became clear that vanadobin can be extracted only from the subpopulation of signet ring cells and not from the morula cells, proving successfully that the substance is not an artifact from the preparation (Michibata *et al.*, submitted). Study of the chemical structure of the vanadobin is in progress.

7. The Valency of Vanadium in the Ascidians

In living organisms, vanadium is present in the $+5$, $+4$, and $+3$ oxidation states. The $+3$ oxidation state of vanadium is strongly reduced($VO^{2+} + 2H^+ + e^- = V^{3+} + H_2O$, $E_0 = 0.36V$) and thus its presence in a living system is unusual. Henze first suggested that the yellow-green blood cells of *Phallusia mammillata* contained vanadium in $+5$ oxidation state. Later, on the basis of oxidative potentiometric measurements (Boeri, 1952) he suggested the occurrence of the $+2$ oxidation state of vanadium, which is more reducing than the $+3$ state ($V^{3+} + e^- = V^{2+}$, $E_0 = -0.26V$) (Henze, 1932). Bielig *et al.* (1954) reported that the red-brown component in the hemolysate (Henze solution) of blood cells of *P.*

mammillata was a protein bound species with vanadium in the + 3 oxidation state in sulfuric acid. Lybing (1953) and Webb (1956) also found the + 3 oxidation state of vanadium. Rezayeva (1964) proposed that there was a dynamic equilibrium of V(III) and V(IV) in the blood cells of *A. aspersa*.

However, these early works were carried out using lysate of blood cells susceptible to air-oxidation and subsequent unknown chemical changes. Thus, non-invasive physical methods, including ESR (electron spin resonance), EXAFS (extended X-ray absorption fine structure), XAS (X-ray absorption spectrometry), SQUID (superconducting quantum interference device) and NMR (nuclear magnetic resonance), enabled the determination of the true intracellular oxidation state of vanadium. Swinehart *et al.* (1974) was the first to use ESR spectroscopy of vanadium in the + 4 oxidation state (VO^{2+}) in ascidians and suggested the presence of vanadium in the + 3 oxidation state in the suborder Phlebobranchia. Then, Kustin *et al.* (1976) reported that blood extracted in a nitrogen environment showed no ESR signal but air-oxidized, two-week old blood yielded a VO^{2+} signal, indicating that vanadium present in the blood cells is therefore in the + 3 valence state. His group determined the concentration of V(IV) in the cells to be 5.7 mM, representing about 4% of the total vanadium in the blood cells (Dingley *et al.*, 1981). In addition, several important findings on the oxidation state of vanadium were obtained with ESR spectrometry (Bell *et al.*, 1982; Hawkins *et al.*, 1983; Frank *et al.*, 1986; Michibata *et al.*, 1986b; 1987; Brand *et al.*, 1987; Sakurai *et al.*, 1987; 1988).

Proton NMR spectra of living blood cells of *A. ceratodes* was found to exhibit a broad 21 ppm signal which corresponded to a labile vanadium(III) aquo complex (Carlson, 1975). Analysis of XAS data of either living or spontaneously lysed blood cells of *A. ceratodes* showed that less than 10% of the vanadium was present as VO^{2+} the remaining being as V^{3+}; EXAFS data gave no evidence for the presence of VO^{2+} (Tullius *et al.*, 1980). Consequently, Tullius *et al.* (1980) proposed a simple aquo-vanadium(III) complex in ascidian blood cells (Figure 6). With a SQUID susceptometer, Lee *et al.* (1988) obtained consistent results for vanadium(III) being the predominant species in both whole and freeze-dried blood sample of *A. nigra* (Figure 7). Recent analysis of the blood cells of *A. ahodori* by ESR revealed that the vanadyl form was not over 10% (Sakurai *et al.*, in preparation).

8. Mechanism of Accumulation of Vanadium in the Blood Cells

Sea water is reported to contain dissolved vanadium at a concentration of 35 nM (Cole *et al.*, 1983; Collier, 1984) but concentrations as high as 100 mM have been found in ascidian blood cells (Macara *et al.*, 1979b; Michibata *et al.*, 1986a), where the concentration of vanadium is in excess of one million times that in sea water. The question of the vanadium oxidation state in sea water is still experimentally unresolved (Biggs and Swinehart, 1976). Although it seems generally that the oxidation state of vanadium in sea water is + 5 as HVO_4^{2-} or $H_2VO_4^-$ (McLeod

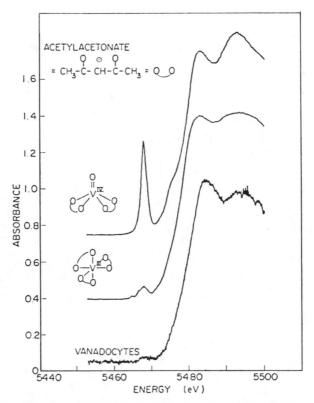

Fig. 6. Vanadium edge spectra of VO(acac)$_2$, V(acac)$_3$ and vanadocytes. The vanadocytes edge spectrum is very similar to the V(acac)$_3$ spectrum, indicating a symmetrical coordination environment for vanadium in the +3 oxidation state in the living vanadocytes (Tullius *et al.*, 1980).

et al., 1975). Sugimura *et al.* (1978) reported that 80 to 89% of total vanadium in sea water is present as organic forms.

Goldberg *et al.* (1951) first studied mechanism of vanadium uptake by ascidians using [48]V radioisotope and demonstrate that whole animals of *A. ceratodes* and *Ciona intestinalis* were capable to concentrating radioactive vanadate directly from sea water by an absorption mechanism and that the uptake was inhibited by the addition of phosphate. Rummel *et al.* (1966) investigated more precisely the accumulation mechanism of vanadium by the ascidian, *C. intestinalis*, with [48]V. The branchial basket yielded more than half of the incorporated vanadium and the gastrointestinal tract contained 10%. The uptake of [48]V into the branchial basket was highly temperature dependent, and could be inhibited by ouabain. In addition, the vanadium bound with blood plasma was transferred to the blood cells 5 days after the labeling.

Using a combination technique of [48]V and ESR spectrometry, Kustin's group, attempted to clarify the accumulation mechanism of vanadium by the blood cells.

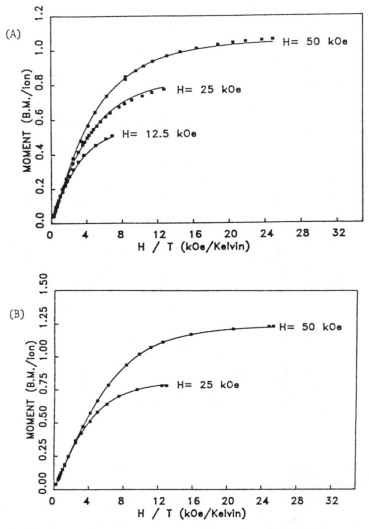

Fig. 7. Magnetic moment per ion versus H/T (H: applied magnetic field intensity and T: absolute temperature (K)) of whole *Ascidia nigra* blood cells (A) and vanadium(III) in solution, V(catechol)$_3$$^{2-}$ ion (B) (Lee *et al.*, 1989).

Dingley *et al.* (1981) observed that the influx of vanadate to tne blood cells of *A. nigra* was rapid (t$_{1/2}$ = 57 sec at 0 °C) with monophasic process. During the influx of vanadate(V) to the blood cells, the ESR signal of intracellular vanadyl(IV) increases, indicating that transported V(V) was reduced upon entering the cell and then as the rate of influx slows, the amount of newly-generated V(IV) in the cells decreased due to further reduction to V(III). Based on these results, they proposed a hypothesis that vanadate enters the blood cells through anionic channels, where it is reduced to the +3 oxidation state. The resulting cations may be trapped as tightly bound complexes, or as free ions which the anionic channel will not accept for transport (Kustin *et al.*, 1983). They postulate that the accumulation is driven by rapid reduction of intracellular vanadate to vanadium in the +4 and/or +3 oxidation states. However, the electron donor for the reduction of vanadate in the vanadocytes is not yet clear.

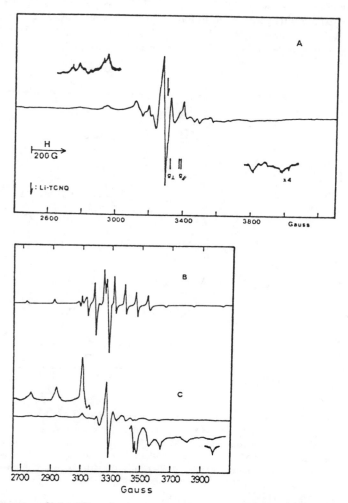

Fig. 8. ESR spectra of vanadyl species in branchial basket washed and treated with the medium containing 100 mM EDTA (A), $VOSO_4$ dissolved in the medium (B), and $VOSO_4$ dissolved in egg albumin solution at pH 7.8 (C). Recording conditions for (A): power, 5 mW; gain, 500; modulation, 6.3 G; frequency, 9.2080; and temperature of 77 K. Similar conditions were used for samples (B) and (C) (Sakurai *et al.*, 1988).

The determination of vanadium in several tissues of ascidians by means of a neutron activation analysis revealed that almost all species had the largest vanadium content in their blood cells and secondly in the branchial basket (Michibata *et al.*, 1986a). The ESR spectrum of the branchial basket of *A. ahodori* (Sakurai *et al.*, 1988), is characteristic of a macromolecule-bound vanadyl species such as vanadyl-albumin complex formed at pH 7.8, but not of an aquo-vanadyl complex (Figure 8). The comparison of ESR parameters of vanadyl species detected in the branchial basket with those of a number of model vanadyl complexes indicates that vanadyl ion in the branchial basket ligates tightly with ligands such as hydroxyl, nitrogenous or thiolate groups of amino acid residues in high molecular weight components of the tissue (Figure 9 and Table I). Furthermore, it should be noted that when vanadate ion added to the medium in which pieces

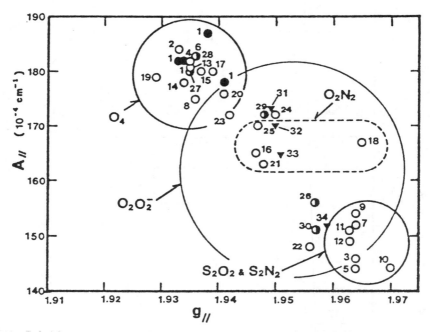

Fig. 9. Relationship between $g_{\parallel}$ and $A_{\parallel}$ for vanadyl species with various ligand fields, $VO(O)_4$, $VO(O_2O_2\text{-})$, $VO(O_2N_2)$, $VO(S_2O_2)$ and $VO(S_2N_2)$ as well as for vanadyl ion detected in branchial baskets of *Ascidia ahodori*. (1) aquavanadyl; (2) VO^{2+}-cysteine at pH 3 and (3) at pH 10; (4) VO^{2+}-cysteine methyl ester at pH 3 and (5) at pH 10; (6) VO^{2+}-mercaptoethylamine at pH 3 and (7) at pH 10; (8) VO^{2+}-glutathione at pH 3 and (9) at pH 10; (10) VO^{2+}-mercaptoethanol at pH 10; (11) VO^{2+}-dithiothreitol at pH 10; (12) VO^{2+}-mercaptopropionic acid at pH 10; (13) VO^{2+-}-ascorbic acid at pH 3.5; (14) VO^{2+}-catechol at pH 6.0; (15) VO^{2+}-serine at pH 3 and (16) at pH 10.5; (17) VO^{2+}-threonine at pH 4.0 and (18) at pH 10.5; (19) VO^{2+}(ATP) at pH 3.0, (20) at pH 7.7, (21) at pH 8.1 and (22) at pH 12.1; (23) supernatant of rat liver; (24) mitochondria of rat liver; (25) radish leaves treated with sodium vanadate; (26) *Phallusia julinea*; (27) *Ascidia ceratodes*; (28) *A. sydneiensis samea*; (29) *A. ahodori* (Asamushi, Aomori, Japan); (30) *A. ahodori* (Ushimado, Okayama, Japan); (31-34)(these numbers refer to Table I) (Sakurai *et al.*, 1988).

Table I. ESR parameters[a] of vanadyl ions detected in branchial basket of *A. ahodori*.

System	g_0	$g_\parallel$	$g_\perp$	A_0	$A_\parallel$ (Gauss/10^{-4} cm^{-1})	$A_\perp$	No.[b]
Supernatant of washed	1.976	1.949	1.989	107	190	66	31
branchial basket				(99)	(173)	(61	)
Branchial basket (washed)	1.976	1.950	1.989	106	187	65	32
				(98)	(170)	(60	)
Branchial basket	1.979	1.951	1.993	101	182	61	33
treated with 100 mM				(96)	(165)	(57	)
EDTA	1.981	1.959	1.993	93	166	61	34
				(89)	(151)	(57	)
Branchial basket	1.977	1.951	1.990	103	182	64	
treated with 100 mM				(99)	(168)	(59	)
EDTA + NaVO$_3$	1.979	1.959	1.990	95	166	64	
(after 5 h)				(91)	(153)	(59	)

[a] To g_0 and A_0 values were calculated by the equations $g_0 = (g_\parallel + 2g_\perp)/3$ and $A_0 = (A_\parallel + 2A_\perp)/3$, respectively, using sets of $g_\parallel$ and $g_\perp$ values and $A_\parallel$ and $A_\perp$ values, obtained from the ESR spectra in the frozen (77 K) state.
[b] The numbers refer to Fig. 3. [Sakurai *et al.*, 1988].

of branchial basket were immersed, the signal height due to vanadyl ion increased approximately two-fold after 5 hr at 10 °C, indicating that the added vanadate ion was reduced to vanadyl ion by the branchial basket. These findings suggests that vanadate(V) incorporated into ascidians is at first reduced to vanadyl(IV) species in the branchial basket, and then the reduced vanadyl is transferred to the vanadocytes, where it will be further reduced to vanadic(III) forms and accumulated.

9. Problem of Acidity in the Blood Cells

The mechanism of accumulation of high levels of vanadium in ascidian blood cells has been considered to be associated with a very low pH value within the cells because the redox potential of the intracellular components is pH dependent and because the type of vanadium species that would exist in the cells is pH dependent (Boeri, 1952). The problem of intracellular pH of vanadocytes is, however, in utter confusion at present. Following Henze's discovery of 1 N acidity in the blood cells of *Phallusia mammillata* (Henze, 1911), 1.83 N sulfuric acid and 0.39 N acid were measured by titration method of the acid liberated upon cytolysis of the blood cells of *P. mammillata* (Webb, 1939) and *Pyura stolonifera* (Endean, 1955a), respectively.

Kustin's group (Dingley *et al.*, 1982; Agudelo *et al.*, 1983) recently claimed that the methods used previously gave spurious results when applied to ascidian blood cells. The cell interior, including high levels of vanadium and iron, is very probably a highly reducing environment and therefore the intracellular redox potential and not the pH was very likely measured in earlier work. They reported that the intracellular pH was neutral on the basis of measurements made by a new technique with improved trans-membrane equilibrium of ^{14}C-labeled methylamine.

Hawkins' group (Hawkins *et al.*, 1983; Brand *et al.*, 1987), has pointed out that the above method suffers from the problem that the distribution of the bases across the membrane may be a function of the complexation of the bases with intracellular metals, and not simply a function of the pH-gradient across the membrane. Thus, they used a non-invasive probe based on the chemical shift of ^{31}P-NMR and reported that ^{31}P-NMR spectrum of blood cells of *Pyura stolonifera* had Pi signals corresponding to pH 6.4 and 6.1, probably from phosphate within the globules and cytoplasm of the morula cells. Conversely, Frank *et al.* (1986) reported a blood cell pH value of 1.8 based on a new finding that the ESR line width reflected accurately the intracellular pH, a method which is also non-invasive.

However, these previous studies focused on the greenish-hued morula cells, considered to be vanadocytes. But these works, of course, did not target the signet ring cells, newly identified as the vanadocytes. We suppose that one or two types of blood cell among several types must have a highly acidic solution within their vacuoles where the vanadium ions would be present in a reduced state. Therefore, one of the reasons for the variation in pH values reported for ascidian blood cells is probably due to the measurement of pH without cell fractionation. In fact, the combined technique involving cell fractionation by density gradient centrifugation on Ficoll type 400, a microelectrode for pH measurement and neutron activation analysis for vanadium determination may reveal that the signet ring cells of *A. ahodori* had a low pH of 2.6 under both aerobic and anaerobic conditions and contained a high amount of vanadium (Michibata *et al.*, in preparation). Since sulfur coexists with vanadium in ascidian blood cells (Bell *et al.*, 1982; Scippa *et al.*, 1985; 1988), it is possible that the accumulation and reduction of vanadium ion is carried out in the presence of sulfuric acid in the blood cells.

10. Function of Vanadium in the Ascidians

The function of such a high concentration of vanadium in ascidian blood cells remains unexplained. Endean (1955a; 1955b; 1955c; 1960) and Smith (1970a; 1970b) proposed that the cellulose of the tunic might be produced by vanadocytes. Carlisle (1968) has suggested that vanadium-containing vanadocytes can reversibly trap oxygen under conditions of low oxygen tension. Swinehart *et al.* (1974) suggested that the acid-producing function of vanadium, and Rowley (1983) has considered that the vanadium contained in vacuole worked as an antimicrobial agent. However, almost all of the proposals put forward have recently become doubtful on the basis of newer experimental evidence or have won unquestioned acceptance. We maintain that any proposals concerning the function of vanadium in the ascidians should be reconsidered in the light of current experimental evidence.

Vanadium exists in several oxidation states and easily forms coordination complexes (Chasteen et al., 1981; 1983). Therefore, in living organisms other than ascidians, vanadium has been shown to be involved in enzymatic and redox reactions. For example, vanadium functions as an active center in bromoperoxidase, extracted from marine algae (de Boer et al., 1986a; 1986b), and in nitrogenase purified from bacteria (Robson et al., 1986; Morningstar et al., 1987; Kentemich et al., 1988). A vanadium complex, named amavadin, was isolated from fly agaric (Bayer et al., 1987). Vanadate is a potent inhibitor of $(Na^+ + K^+)$-ATPase (Cantley et al., 1977; Beauge and Glynn, 1977). Recently, it has become clear that vanadium exerts an insulin-like effect on diabetic rats (Dubyak and Kleinzeller, 1980; Heyliger et al., 1985) and an inhibitory effect on carcinogenesis (Thompson et al., 1984). Resolution of the above issues in the ascidians would help to clarify the function of transition metals in living organisms other than ascidians.

11. Conclusion

The fact that some ascidians accumulate vanadium ion specifically to a level exceeding four million times that present in sea water is an unusual phenomenon and the fact that they are the only organisms in animal kingdom to accumulate vanadium at such a high level is a mystery. Interdisciplinary investigations, based on marine biology, physiology, biochemistry, bioinorganic chemistry and the chemistry of natural products, are absolutely necessary for understanding these phenomena (Michibata, 1989).

Acknowledgments

We would like to express our hearty thanks to our associates for their friendly assistance. Particular thanks are due to the staff of marine biological stations of the Asamushi, Tohoku University, the Ushimado, Okayama University, and the Misaki and the Ootsuchi, University of Tokyo. This work was supported in part by a Grant-in-Aid from the Ministry of Education, Science and Culture, Japan to H.M. (Nos. 61030035, 62540540, 01304007 and 01480026) and was also supported financially by the Japan Securities Scholarship Foundation, the Ito Science Foundation, and the Tamura Foundation for the Promotion of Science and Technology. Neutron activation analysis was carried out under the Cooperative Programs of the Institute for Atomic Energy of Rikkyo University.

References

Agudelo, M. I., Kustin, K. and McLeod, G. C. (1983) The intracellular pH of the blood cells of the tunicate Boltenia ovifera. Comp. Biochem. Physiol., 75A: 211-214.

Bayer, E., Koch, E. and Anderegg, G. (1987) Amavadin, an example for selective binding of vanadium in nature: Studies of its complexation chemistry and a new structural proposal. Angew. Chem. Int. Ed. Engl., 26: 545-546.

Beauge, L. A. and Glynn, I. M. (1977) A modifier of (Na + + K +)ATPase in commercial ATP. Nature, 268: 355-356.

Bell, M. V., Pirie, B. J. S., McPhail, D. B., Goodman, B. A., Falk-Petersen, I.-B. and Sargent, J. R. (1982) Contents of vanadium and sulphur in the blood cells of Ascidia mentula and Ascidiella aspersa. J. Mar. Biol. Ass. U.K., 62: 709-716.

Bertrand, D. (1950) Survey of contemporary knowledge of biogeochemistry. 2. The biogeochemistry of vanadium. Bull. Am. Museum Natl. History, 94: 403-455.

Bielig, H. -J., Bayer, E., Califano, L. and Wirth, L. (1954) The vanadium-containing blood pigment. II. Hemovanadin, a sulfate complex of trivalent vanadium. Publ. Staz. Zool. Napoli, 25: 26-66.

Bielig, H. -J., Jost, E., Pfleger, K., Rummel, W. and Seifen, E. (1961) Aufnahme und Verteilung von Vanadin bei der Tunicate Phallusia mammillata Cuvier (Untersuchungen uber Hamovanadin, V). Hoppe-Seyer's Z. Physiol. Chem., 325: 122-131.

Bielig, H. -J., Bayer, E., Dell, H. -D., Rohns, G., Mollinger, H. and Rudiger, W. (1966)Chemistry of hemovanadin. Protides Biol. Fluids, 14: 197-204.

Biggs, W. R. and Swinehart, J. H. (1976) Vanadium in selected biological systems. In "Metal ions in biological systems. Vol. 6, Biological action of metal ions." Edit.by Sigel, H., Marcel Dekker, Inc., New York., pp141-196.

Boeri, E. (1952) The determination of hemovanadin and its oxidation potential. Arch. Biochem. Biophys., 37: 449-456.

Boeri, E. and Ehrenberg, A. (1954) On the nature of vanadium in vanadocyte hemolysate from ascidians. Arch. Biochem. Biophys., 50: 404-416.

Botte, L. and Scippa, S. (1977) Ultrastructural study of vanadocytes in Ascidia malaca. Experientia, 33: 80-81.

Botte, L. and Scippa, S. (1979) Vanadium and other metals in Ascidia malaca. Annot. Zool. Japon., 52: 188-190.

Botte, L., Scippa, S. and de Vincentiis, M. (1979a) Content and ultrastructural localization of transitional metals in ascidian ovary. Develop. Growth Differ., 21: 483-491.

Botte, L., Scippa, S. and de Vincentiis, M. (1979b) Ultrastructural localization of vanadium in the blood cells of Ascidiacea. Experientia, 35: 1228-1230.

Brand, S. G., Hawkins, C. J. and Parry, D. L. (1987) Acidity and vanadium coordination in vanadocytes. Inorg. Chem., 26: 627-629.

Bruening, R. C., Oltz, E. M., Furukawa, J., Nakanishi, K. and Kustin, K. (1985) Isolation and structure of tunichrome B-1, a reducing blood pigment from the tunicate Ascidia nigra L. J. Am. Chem. Soc., 107: 5298-5300.

Califano, L. and Boeri, E. (1950) Studies on haemovanadin. III. Some physiological properties of haemovanadin, the vanadium compound of the blood of Phallusia mammillata Cuv. J. Exp. Zool., 27: 253-256.

Cantacuzene, J. and Tchekirian, A. (1932) Sur la presence de vanadium chez certains tuniciers. Compt. Rend. Acad. Sci. Paris, 195: 846-849.

Cantley, L. C. Jr., Josephson, L., Warner, R., Yanagisawa, M., Lechene, C. and Guidotti, G. (1977) Vanadate is a potent (Na, K)-ATPase inhibitor found in ATP derived from muscle. J. Biol. Chem., 252: 7421-7423.

Carlisle, D. B. (1958) Niobium in ascidians. Nature, 181: 933.

Carlisle, D. B. (1968) Vanadium and other metals in ascidians. Proc. Roy. Soc., B, 171: 31-42.

Carlson, R. M. K. (1975) Nuclear magnetic resonance spectrum of living tunicate blood cells and structure of the native vanadium chromogen. Proc. Nat. Acad. Sci. USA, 72: 2217-2221.

Chasteen, N. D. (1981) Vanadyl(IV) EPR spin probes. Inorganic and biochemical aspects. In "Biological magnetic resonance." Vol. 3, Edit. Berliner, L. and Reuben, J. Plenum Press, New York, pp 53-119.

Chasteen, N. D. (1983) The biochemistry of vanadium. Structure and Bonding, 53: 105-138.

Ciereszko, L. S., Ciereszko, E. M., Harris, E. R. and Lane, C. A. (1963) Vanadium content of some tunicates. Comp. Biochem. Physiol., 8: 137-140.

Cole, P. C., Eckert, J. M. and Williams, K. L. (1983) The determination of dissolved and particulate vanadium in sea water by X-ray fluorescence spectrometry. Anal. Chim. Acta, 153: 61-67.

Collier, R. W. (1984) Particulate and dissolved vanadium in the North Pacific Ocean. Nature, 309: 441-444.

Danskin, G. P. (1978) Accumulation of heavy metals by some solitary tunicates. Can. J. Zool., 56: 547-551.

de Boer, E., Tromp. M. G. M., Plat, H., Krenn, G. E. and Wever, R. (1986a) Vanadium(V) as an essential element for haloperoxidase activity in marine algae: Purification and characterization of a vanadium-containing bromoperoxidase from Laminaria saccharina. Biochim. Biophys. Acta, 872: 104-115.

de Boer, E., van Kooyk, Y., Tromp, M. G. M., Plat, H. and Wever, R. (1986b) Bromoperoxidase from Ascophyllum nodosum: A novel class of enzymes containing vanadium a prosthetic group? Biochim. Biophys. Acta, 869: 48-53.

de Vincentiis, M. (1962) Ulteriori indagini istospettrografiche e citochimiche su alcuni aspetti dell'ovogenesi di Ciona intestinalis. Atti Soc. Peloritana, Sc. Fis. Mat. Nat., 8: 189-198.

de Vincentiis, M. and Rudiger, W. (1967) Fluorescence of blood cell of the tunicates Phallusia mammillata and Ciona intestinalis. Experientia, 23: 245-246.

Deno, T. (1987) Autonomous fluorescence of eggs of the ascidian Ciona intestinalis. J. Exp. Zool., 241: 71-79.

Dingley, A. l., Kustin, K., Macara, I. G. and McLeod, G. C. (1981) Accumulation of vanadium by tunicate blood cells occurs via a specific anion transport system. Biochim. Biophys. Acta, 649: 493-502.

Dingley, A. L., Kustin, K., Macara, I. G., McLeod, G. C. and Roberts, M. F. (1982) Vanadium-containing tunicate blood cells are not highly acidic. Biochim. Biophys. Acta, 720: 384-389.

Dubyak, G. R. and Kleinzeller, A. (1980) The insulin-mimic effects of vanadate in isolated rat adipocytes. J. Biol. Chem., 255: 5306-5312.

Endean, R. (1955a) Studies of the blood and tests of some Australian ascidians. I. The blood of Pyura stolonifera (Heller). Austr. J. Mar. Freshwat. Res., 6: 35-59.

Endean, R. (1955b) Studies of the blood and tests of some Australian ascidians. II. The test of Pyura stolonifera (Heller). Austr. J. Mar. Freshwat. Res., 6: 139-155.

Endean, R. (1955c) Studies of the blood and tests of some Australian ascidians. III. The formation of the test of Pyura stolonifera (Heller). Austr. J. Mar. Freshwat. Res., 6: 157-164.

Endean, R. (1960) The blood-cells of the ascidian, Phallusia mammillata. Quart. J. Microscop. Sci., 101: 177-197.

Ermak, T. H. (1975) Autoradiographic demonstration of blood cell renewal in Styela clava (Urochordata: Ascidiacea). Experientia, 31: 837-839.

Frank, P., Carlson, R. M. K. and Hodgson, K. O. (1986) Vanadyl ion EPR as a noninvasive probe of pH in intact vanadocytes from Ascidia ceratodes. Inorg. Chem., 25: 470-478.

Freeman, G. (1964) The role of blood cells in the process of asexual reproduction in the tunicate Perophora viridis. J. Exp. Zool., 156: 157-184.

Gilbert, K., Kustin, K. and McLeod, G. C. (1977) Gel filtration analysis of vanadium in Ascidia nigra blood cell lysate. J. Cell. Physiol., 93: 309-312.

Goldberg, E. D., McBlair, W. and Taylor, K. M. (1951) The uptake of vanadium by tunicates. Biol. Bull., 101: 84-94.

Goodbody, I. (1974) The physiology of ascidians. Adv. Mar. Biol., 12: 1-149.

Hawkins, C. J., Kott, P., Parry, D. and Swinehart, J. H. (1983) Vanadium content and oxidation state related to ascidian phylogeny. Comp. Biochem. Physiol., 76B: 555-558.

Henze, M. (1911) Untersuchungen uber das Blut der Ascidien. I. Mitteilung. Die Vanadium-verbinding der Blutkorchen. Hoppe-Seyer's Z. Physiol. Chem., 72: 494-501.

Henze, M. (1932) Uber das Vanadium Chromogen des Ascidienblutes. Hoppe-Seyer's Z. Physiol. Chem., 213: 125-135.

Heyliger, C. E., Tahiliani, A. G. and McNeill, J. H. (1985) Effect of vanadate on elevated blood glucose and depressed cardiac performance of diabetic rats. Science, 277: 1474-1477.

Kalk, M. (1963a) Absorption of vanadium by tunicates. Nature, 198: 1010-1011.

Kalk, M. (1963b) Intracellular sites of acidity in the histogenesis of tunicate vanadocytes. Quart. J. Microsc. Sci., 104: 483-493.

Kalk, M. (1963c) Cytoplasmic transmission of a vanadium compound in a tunicate oocyte, visible with electronmicroscopy. Acta Embryol. Morph. Exp., 6: 289-303.

Kentemich, T., Danneberg, G., Hundeshagen, B. and Bothe, H. (1988) Evidence for the occurrence of the alternative, vanadium-containing nitrogenase in the cyanobacterium *Anabaena variabilis*. FEMS Microbiol. Let., 51: 19-24.

Kobayashi, S. (1935) On the presence of vanadium in certain Pacific ascidians. Sci. Rep. Tohoku Univ. 4th Ser., 18: 185-193.

Kokubu, N. and Hidaka, T. (1965) Tantalum and niobium in ascidians. Nature, 205: 1028-1029.

Kustin, K., Levine, D. S., McLeod, G. C. and Curby, W. A. (1976) The blood of Ascidia nigra: Blood cell frequency distribution, morphology, and the distribution and valence of vanadium in living blood cells. Biol. Bull., 150: 426-441.

Kustin, K., McLeod, G. C., Gilbert, T. R. and Briggs, L.B.R. 4th (1983) Vanadium and other metal ions in the physiological ecology of marine organisms. Structure and Bonding, 53: 139-160.

Lee, S., Kustin, K., Robinson, W. E., Frankel, R. B. and Spartalian, K. (1988) Magnetic properties of tunicate blood cells. I. *Ascidia nigra*. J. Inorg. Biochem., 33: 183-192.

Levine, E. P. (1961) Occurrence of titanium, vanadium, chromium, and sulfuric acid in the ascidian *Eudistoma ritteri*. Science, 133: 1352-1353.

Lybing, S. (1953) The valence of vanadium in hemolysates of blood cells from *Ascidia obliqua* Alder. Arkiv Kemi., 6: 261-269.

Macara, I. G. (1980) Vanadium – an element in search of a role. Trends Biochem. Sci., 5: 92-94.

Macara, I. G., McLeod, G. C. and Kustin, K. (1979a) Isolation, properties and structural studies on a compound from tunicate blood cells that may be involved in vanadium accumulation. Biochem. J., 181: 457-465.

Macara, I. G., McLeod, G. C. and Kustin, K. (1979b) Tunichromes and metal ion accumulation in tunicate blood cells. Comp. Biochem. Physiol., 63B: 299-302.

McLeod, G. C., Ladd, K. V., Kustin, K. and Toppen, D.l. (1975) Extraction of vanadium(V) from seawater by tunicates: A revision of concepts. Limnol. Oceanogr., 20: 491-493.

Michibata, H. (1984) Comparative study on amounts of trace elements in the solitary ascidians, *Ciona intestinalis* and *Ciona robusta*. Comp. Biochem. Physiol., 78A: 285-288.

Michibata, H. (1989) New aspects of accumulation and reduction of vanadium ions in ascidians, based on concerted investigation for both a chemical and biological viewpoint. Zool. Sci., 6: 639-647.

Michibata, H., Terada, T., Anada, N., Yamakawa, K. and Numakunai, T. (1986a) The accumulation and distribution of vanadium, iron, and manganese in some solitary ascidians. Biol. Bull., 171: 672-681.

Michibata, H. Miyamoto, T. and Sakurai, H., (1986b) Purification of vanadium binding substance from the blood cells of the tunicate, *Ascidia sydneiensis samea*. Biochem. Biophys. Res. Commun., 141: 251-257.

Michibata, H., Hirata, J., Uesaka, M., Numakunai, T. and Sakurai, H. (1987) Separation of vanado-cytes: Determination and characterization of vanadium ion in the separated blood cells of the ascidian, *Ascidia ahodori*. J. Exp. Zool., 244: 33-38.

Michibata, H., Hirata, J., Terada, T. and Sakurai, H. (1988) Autonomous fluorescence of ascidian blood cells with special reference to identification of vanadocytes. Experientia, 44: 906-907.

Michibata, H., Uyama, T. and Hirata, J. (1989a) Vanadium-containing blood cells (vanadocytes) show no fluorescence due to the tunichrome in the ascidian, *Ascidia sydneiensis samea*. Zool. Sci., in press.

Michibata, H. and Uyama, T. (1989b) Extraction of vanadium binding substance (vanadobin) from a subpopulation of signet ring cells newly identified as vanadocytes in ascidians. J. Exp. Zool., in press.

Morningstar, J. E., Johnson, M. K., Case, E. E. and Hales, B. J. (1987) Characterization of the metal cluster in the nitrogenase molybdenum-iron and vanadium-iron proteins of *Azotobacter vinelandii* using magnetic circular dichroism spectroscopy. Biochemistry, 26: 1795-1800.

Nishide, K., Nishikata, T. and Satoh, N. (1989) A monoclonal antibody specific to embryonic trunk-lateral cells of the ascidian *Halocynthia roretzi* stains coelomic cells of juvenile and adult basophilic blood cells. Develop. Growth Differ., in press.

Noddack, I. and Noddack, W. (1939) Die Haufigkeiten der Schwermetalle in Meerestieren. Arkiv. Zool., 32: 1-35.

Rezayeva, L. T. (1964) Valency state of vanadium in blood cells of *Ascidiella aspersa*. Zhurnal Obshchei Biologii, 25: 374-378.

Robinson, W. E., Agudelo, M. I. and Kustin, K. (1984) Tunichrome content in the blood cells of the tunicate, *Ascidia callosa* Stimpson, as an indicator of vanadium distribution. Comp. Biochem. Physiol., 78A: 667-673.

Robson, R. L., Eady, R. R., Richardson, T. H., Miller, R. W., Hawkins, M. and Postgate, J. R. (1986) The alternative nitrogenase of *Azotobacter chroococcum* is a vanadium enzyme. Nature, 322: 388-390.

Roman, D. A., Molina, J. and Rivera, L. (1988) Inorganic aspects of the blood chemistry of ascidians. Ionic composition, and Ti, V, and Fe in the blood plasma of *Pyura chilensis* and *Ascidia dispar*. Biol. Bull., 175: 154-166.

Rowley, A. F. (1982) the blood cell of *Ciona intestinalis*: An electron probe X-ray microanalytical study. J. Mar. Biol. Ass. U.K., 62: 607-620.

Rowley, A. F. (1983) Preliminary investigations on the possible antimicrobial properties of tunicate blood cell vanadium. J. Exp. Zool., 227: 319-322.

Rummel, W., Bielig, H. -J., Forth, W., Pfleger, K., Rudiger, W. and Seifen, E. (1966) Absorption and accumulation of vanadium by tunicates. Protides Biol. Fluids, 14: 205-210.

Sakurai, H., Hirata, J. and Michibata, H. (1987) EPR characterization of vanadyl(IV) species in blood cells of ascidians. Biochem. Biophys. Res. Commun., 149: 411-416.

Sakurai, H., Hirata, J. and Michibata, H. (1988) ESR spectra of vanadyl ion detected in branchial basket of an ascidian, *Ascidia ahodori*. Inorg. Chim. Acta, 152: 177-180.

Scippa, S., Botte, L. and de Vincentiis, M. (1982) Ultrastructure and X-ray microanalysis of blood cells of *Ascidia malaca*. Acta Zool. (Stokh.), 63: 121-131.

Scippa, S., Botte, L., Zierold, K. and de Vincentiis, M. (1985) X-ray microanalytical studies on cryofixed blood cells of the ascidian *Phallusia mammillata*. Cell Tissue Res., 239: 459-461.

Scippa, S., Zierold, K. and de Vincentiis, M. (1988) X-ray microanalytical studies on cryofixed blood cells of the ascidian *Phallusia mammillata*. II. Elemental composition of the various blood cell types. J. Submicrosc. Cytol. Pathol., 20: 719-730.

Smith, M. J. (1970a) The blood cells and tunic of the ascidian *Halocynthia aurantium* (Pallas). I. Hematology, tunic morphology, and partition of cells between blood and tunic. Biol. Bull., 138: 354-378.

Smith, M. J. (1970b) The blood cells and tunic of the ascidian *Halocynthia aurantium* (Pallas). II. Histochemistry of the blood cells and tunic. Biol. Bull., 138: 379-388.

Smith, M. J. (1989) Vanadium biochemistry: The unknown role of vanadium-containing cells in ascidians (sea squirts). Experientia, 45: 452-457.

Sugimura, Y., Suzuki, Y. and Miyake, Y. (1978) Chemical forms of minor metallic elements in the ocean. J. Oceanogr. Soc. Japan, 34: 93-96.

Swinehart, J. H., Biggs, W. R., Halko, D. J. and Schroeder, N. C. (1974) The vanadium and selected metal contents of some ascidians. Biol. Bull., 146: 302-312.

Thompson, H. J., Chasteen, N. D. and Meeker, L. D. (1984) Dietary vanadyl(IV) sulfate inhibits chemically-induced mammary carcinogenesis. Carcinogenesis, 5: 849-851.

Tullius, T. D., Gillum, W. O., Carlson, R. M. K. and Hodgson, K. O. (1980) Structural study of the vanadium complex in living ascidian blood cells by X-ray absorption spectrometry. J. Am. Chem. Soc., 102: 5670-5676.

Vinogradov, A. P. (1934) Distribution of vanadium in organism. C. R. Acad. Sci. U.R.S.S., 3: 454-459.

Webb, D. A. (1939) Observations on the blood of certain ascidians, with special reference to the biochemistry of vanadium. J. Exp. Biol., 16: 499-523.

Webb, D. A. (1956) The blood of tunicates and the biochemistry of vanadium. Publ. Staz. Zool. Napoli, 28: 273-288.

Wright, R. K. (1981) Urochordates. In "Invertebrate blood cells." Vol. 2. Edit. by Ratcliffe, N. A. and Rowley, A. F., Academic Press, London, pp. 565-626.

X. Biological Applications of ^{51}V NMR Spectroscopy

DIETER REHDER

Institute of Inorganic Chemistry, The University, Martin Luther King Platz 6, D-2 Hamburg 13, F.R.G.

1. Introduction

Application of ^{51}V NMR to genuine biological systems has developed only during the last few years and consequently is still restricted to a small number of examples, viz. vanadate binding to serum albumin and apo-transferrin, the nature of the coordination environment of vanadium in vanadate-dependent halo-peroxidases from knobbed wrack (a marine brown alga), and the interaction of vanadate with several enzymes involved in phosphorylation reactions (phospho-glycerate mutase, alkaline phosphatase, ribonuclease A and ribonuclease T_1). The main problem with *large* biological molecules arises from the quadrupole moment of ^{51}V which, in most cases so far investigated, induces very broad resonance lines (see below). On the other hand, ^{51}V NMR has frequently and successfully been employed to investigate the formation of complexes between vanadate, VO^{3+} or VO_2^+ and small organic molecules, which are part of or closely related to large biological matrices, such as oligopeptides, hydroxycarboxylic acids, phenols, saccharides, functionalized amines, nucleotides and nucleosides. Studies of these systems have revealed insight not only into the ligand behavior of biogenic and bio-mimetic molecules, but also into thermodynamic (and kinetic) properties of simple and complex systems underlying chemical exchange. Examples are the determination of formation constants for the weak complexes formed between vanadate and dipeptides, or esters and mixed anhydides of the hypothetical orthovanadium acid.

^{51}V NMR is in fact a feasible method for investigating the speciation of vanadium under physiological conditions in the presence of inorganic (H_3O^+, O_2^{2-}, phosphate) and organic matter, to which it becomes associated. In addition to an overview on the involvement of vanadium in complex formation with organic ligands, as revealed by ^{51}V NMR, a brief description of inorganic speciation (protonation equilibria, peroxovanadates, iso- and phosphovanadate anions) is also provided. Phosphovanadates are of interest in the context of the vanadate/phosphate antagonism in all living organisms. Peroxovanadates have been included because of their possible relevance to vanadate-dependent peroxidase. Molybdovanadates, which have also been studied by ^{51}V NMR (Maksimovskaya and Chumachenko, 1987; Howarth *et al.*, 1989), and which might be considered interesting species with respect to the competitive use of vanadate and molybdate in nitrogen fixation, have been omitted since, at physiological pH and at the very

N. Dennis Chasteen (ed.), Vanadium in Biological Systems, 173–197.

low concentrations of the ultra trace elements V and Mo, their formation is strongly disfavored.

The ^{51}V nucleus is among the most suitable nuclei for the an NMR experiment (see below). The detection level for a species can be in the concentration range as low as 5-100 micromole/L, depending, of course, on the nature of the specimen (preferably a small and "spherical" molecule), and depending on the sophistication of instrument and operator, and the time available for a measurement. To achieve a signal-to-noise ratio of 5/1, the measuring time on a 360 MHz instrument (operating at 94.7 MHz for ^{51}V), under standard conditions and for a vanadium concentration of 0.1 mmol/L, amounts to about half-an-hour for tetrahedral $H_2VO_4^-$, and 13 days for a strong vanadate-protein complex of M < 20,000 g/mol. On a high resolution pulsed FT instrument, line widths > ca. 3 kHz are almost always a problem, and in these cases, a wide-line CW device is more suitable for the NMR experiment (it also excludes convolution problems). As will be shown in the last section of this chapter, one may arrive at relatively narrow resonance lines even in a protein complex, as exemplified by vanadate-transferrin.

2. General and Background

In many of its biological functions, vanadium acts in the + 5 (d$^\circ$) state, where it is diamagnetic and usually inaccessible to electron absorption spectroscopy in the visible region. Among the transition metals of biological relevance, vanadium is unique in that the ^{51}V nucleus is easily detectable by NMR. Nuclear properties demonstrating this fact are listed in Table I. The relatively large shielding range for vanadium indicates a high intrinsic sensitivity of ^{51}V towards minor changes in its chemical environment. The quadrupole moment can hamper NMR detection through shorting of spin-spin relaxation times T_2 (and spin-lattice relaxation times T_1), giving rise to broad lines. Rapid relaxation can be a problem when vanadium

Table I. NMR properties of the ^{51}V nucleus.

Natural abundance	99.76%
Nuclear spin	7/2
Nuclear electric quadrupole moment	$-0.043(5) \cdot 10^{-28}$ m^2 [a]
Relative receptivity (constant field, ^{1}H = 1)	0.38
Measuring frequency at 2.349 T (^{1}H = 100 MHz)	26.29 MHz
Standard for referencing chemical shifts	VOCl$_3$ (neat)
Magnetic field for standard at 16.0 MHz	1.492 T
Chemical shift range	+ 1500 to − 2100 ppm
Chemical shift range for V(V) compounds	+ 1500 to − 900 ppm
Typical relaxation times in isotropic media	1–10 ms
Typical quadrupole coupling constants	0.5–7 MHz
Survey articles	Rehder, 1982/84/87

[a] Unkel et al. (1989).

is coordinated to large or bulky molecule and/or in a site of low local symmetry. Since the quadrupole moment is, however, rather small, there are generally no severe problems with spectral resolution even in compounds with symmetries as low as C_s or C_1. The quadrupole interaction in fact provides additional NMR parameters (additional to shielding and scalar coupling) not available from spin 1/2 nuclei.

The experimental parameter connected to quadrupole interaction and directly obtainable from the spectrum in isotropic media is the line width, usually measured as half-width $W_{1/2}$ (the width at half-height of the resonance signal). $W_{1/2}$ is related to the peak-to-peak width W_{pp} of the first derivative of the resonance line by $W_{1/2} = \sqrt{3} W_{pp}$. Hence, scanning the spectrum in the first derivative mode improves resolution! If there are no contributions to the line width stemming from chemical exchange, $W_{1/2}$ and the spin-spin relaxation time T_2 are related to the molecular correlation time τ_c describing the solute/solvent interaction and other molecular parameters by

$$W_{1/2} = (\pi T_2)^{-1} = 2\pi/49 \cdot (NQCC)^2 (1 + \eta^2/3)\tau_c \qquad (1)$$

under extreme narrowing conditions. In eq. (1), NQCC is the nuclear quadrupole coupling constant which describes the interaction ("coupling") of the nuclear quadrupole moment Q with the electric field gradient produced at the nucleus by the molecular environment. η is the asymmetry parameter (zero for axial symmetry). Typical values for NQCC(^{51}V) vary from 0.5 to 7 MHz (Rehder, 1987; Paulsen and Rehder, 1982), correlation times τ_c are in the 1-10 ps range *for small molecules* such as vanadyl esters (Paulsen and Rehder, 1982). Hence, with the measuring frequencies ν_o commonly applied, the extreme narrowing condition $2\pi\nu_o >> 1$ is fulfilled. This condition is not necessarily attained when vanadium is bound to a large molecule such as a protein (section 5). Another complication is that chemical exchange at medium exchange rates can cause exchange broadening of resonance lines, in great excess over broadening effects imparted by the quadrupolar interaction. Whether quadrupole relaxation dominates can be checked by varying the temperature or the viscosity; a decrease in the temperature or an increase in the viscosity leads to a more effective relaxation via an increase of τ_c and hence to broader signals (equation (1)).

Under isotropic conditions, quadrupole perturbation is modulated by the Brownian motion and the only observable quadrupole effect is the line broadening in all molecules where the electric field gradient q at the vanadium nucleus does not approach zero (as is the case, inter alia, in all molecules belonging to cubic point groups). Under partially anisotropic conditions, e.g. in meso-phases (liquid crystals), first order quadrupole perturbation comes in, splitting the resonance line into 7 equidistant components of approximately the same intensity. The first order interaction is probably an important factor whenever vanadium becomes immobilized in a large matrix. First order quadrupole splitting has been observed for monohydrogen-orthovanadate aligned in the mixed lyotropic meso-phase formed by tetradecylammonium dodecanate in water (Tracey and Radley, 1985). For a detailed description of ^{51}V quadrupole interactions under isotropic and aniso-tropic conditions see Paulsen and Rehder, 1982.

An important NMR parameter in the context of applications of ^{51}V NMR to biological systems, and usually the only one discussed at all, is the chemical shift $\delta(^{51}V)$. ^{51}V shielding strongly depends on the elctronegativity (or a closely related quantity such as the polarizability) of the ligand functions in that V(V) compounds, like all d° systems, exhibit the inverse electronegativity dependence. This is shown in Figure 1 for a selected number of complexes containing ligand functions which

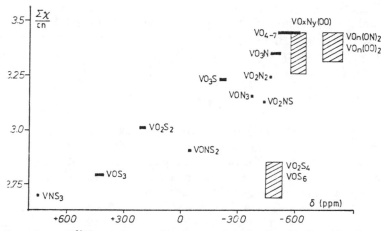

Fig. 1. Dependence of (^{51}V) upon the sum of the electronegativities, Σx, of the ligand functions attached to vanadium, divided by the coordination number cn. Shaded bars indicate complexes containing strained ring structures (O_2^{2-}, $-CO_2^-$, $-CS_2^-$, R_2NO^-). values are given relative to $VOCl_3$. For references to data, cf. Priebsch and Rehder (1985), Rehder et al. (1988).

also occur in nature (S, N, O). The electronegativity criterion may well be used to distinguish between thio ligands on the one hand, and oxo or nitrogen ligands on the other hand. The influences of O- and N-functional ligands upon ^{51}V shielding are quite similar, and a distinction cannot be carried out unambiguously, except where series of very similar compounds are compared.

The inverse electronegativity dependence is a consequence of (i) an increase of the HOMO-LUMO gap and (ii) a decrease of V(3d) contributions to the HOMOs (decrease of covalency) as the ligand electronegativity increases (Figure 1). Super-imposed on electronic effects are steric effects and effects arising from changes in the coordination number and coordination geometry. Generally, an increase of shielding is observed on introducing steric strain via 3- or 4-membered chelate-ring structures (Figure 1), or bulky ligands (inverse dependence on steric factors) (Priebsch and Rehder, 1985; Rehder et al., 1988; Weidemann et al., 1988). The overall increase of the electronegativity of the ligand sphere on increasing the coordination number does not effect $\delta(^{51}V)$ to a great extent (Priebsch and

Rehder, 1985), although there are examples (and these will be discussed in more detail in sections III.1. and III.2.), which demonstrate that the chemical shift can be employed to differentiate between the coordination numbers 4, 5 and 6.

3. The Aqueous Vanadate and Phosphovanadate Systems

A joint emf and ^{51}V NMR study by Pettersson et al. (1983, 1985) of the pH and concentration dependence of the various species present in an aqueous solution when ortho- or metavanadate are dissolved in water (and the pH is subsequently adjusted), has settled the problem of the nature, the conditions for protonation, the relative concentrations, and equilibrium constants for the coexisting mono- and oligovanadates. Other groups have supplemented these investigations by ^{17}O NMR (Harrison and Howarth, 1985) and by specifying the heteropolyvanadates which are formed between vanadate and phosphate (Harrison and Howarth, 1985; Gresser et al., 1986). The latter studies are included here, because of the chemical similarity between vanadate and phosphate, and their ability to form mixed anhydrides, one of the important aspects to be considered in the context of stimulation and inhibition of phosphate-dependent enzymes and enzymatic phosphoryl transfer reactions by vanadate. Further, some of the peroxovanadates detected in the vanadate/H_2O_2/O_2^{2-} system (Harrison and Howarth, 1985a; Campbell et al., 1989) have been included.

Table II summarizes some of the parameters obtained for the species present in solution in the pH range relevant for physiological investigations, i.e. pH 7 $\pm$ 1.5. Figure 2 is a distribution diagram for vanadates at 3 different concentrations. All species are in concentration- and pH-dependent chemical exchange, but only the protonation equilibria and those involving equilibria between phospho-

Table II. Vanadates and phosphovanadates and peroxovanadates present in aqueous solution in the pH range of ca. 5.5–8.5[a].

Vanadate[b]		δ (ppm) [$W_{1/2}$ (Hz)]	$\log(K)$[c] [pK_a]
HV_1	HVO_4^{2-}	-537	
H_2V_1	H_2VO^{4-}	-560 [40–80]	7.9 ($H^+ + HV_1$) [7.9–8.4]
HV_2	$HV_2O_7^{3-}$	-562	9.3 ($H^+ + V_2$) [9.92]
H_2V_2	$H_2V_2O_7^{2-}$	-573 [60–120]	18.6 (2 H^+ + 2 HV_1) [8.02]
V_4	$V_4O_{12}^{4-}$	-576 [50–80]	42 (4 HV_1)
V_5	$V_5O_{15}^{5-}$	-584 [60–80]	52 (5 HV_1)
V_6	$V_6O_{18}^{6-}$	-587	
V_{10}	$V_{10}O_{28}^{6-\ d}$	$-422, -496, -513$	131 (10 HV_1)
HV_{10}	$HV_{10}O_{28}^{5-\ d}$	$-424, -500, -516$	

Table II (continued)

Vanadate[b]		δ (ppm) [$W_{1/2}$ (Hz)]	$\log(K)^c$ [pK_a]
		[530, 330, 200]	[6.00]
HP_1V_1	$HPVO_7^{3-}$	-570 (pH 8.0)	0.76 $(V_1 + P_1)^e$
$H_2P_1V_1$	$H_2PVO_7^{2-}$	-583 (pH 6.7)	1.40 $(V_1 + P_1)^e$
HP_2V_1	$HVP_2O_{10}^{4-}$	-549 (pH 8.0)	1.59 $(H_2V_1 + HP_2)$
$H_2P_4V_1$	$H_2VO_2(P_2O_7)_2^{5-}$	-540 (pH 6.5)	
$VO(OH)(O_2)_2(H_2O)^{2-}$		$-739(2)$ (pH 9.1–4.7) [300]	
$V_2O_3(O_2)_4^{4-}$		-757 (pH 9.1–4.5 [600]	
$VO(O_2)(H_2O)^{-\ f}$		-694 (pH < 6) [300]	

[a] Data for H_xV_1, H_xV_2, V_4, V_5 and V_{10} in 0.6 M NaCl (Petterson *et al.*, 1983) (for K values see footnote c), for V_6 in 3.0 M NaClO$_4$ (Pettersson *et al.*, 1985), for phosphovanadates in 1.0 M KCl (Gresser *et al.*, 1986). δ values of the latter are subject to the molar ratio vanadate/phosphate; δ values for protonated species vary with pH. The shifts indicated are experimentally obtained limiting values. All compounds have tetrahedral site geometry, except for V in $H_2P_4V_1$ (octahedral). A species $HV_3O_{10}^{4-}$ not contained in the Table and invisible as a distinct vanadate in the ^{51}V NMR has also been postulated on the basis of emf studies (Ivakin *et al.*, 1986).
[b] The abbreviations introduced here will be employed throughout.
[c] Formation constant under the ionic strength of 0.6 (Pettersson *et al.*, 1983) or 2.0 M NaCl (Ivakin *et al.*, 1986).
[d] The three signals in the intensity ratio 2/4/4 correspond to 3 different vanadium sites.
[e] The subscript "i" (= inorganic) indicates unspecified monovanadate and monophosphate. $\log(K)$ for the formation of $H_2P_1V_1$ from H_2V_1 and H_2P is 1.81 (K = 64 $\pm$ 3 M^{-1}).
[f] Assignment by Campbell *et al.* (1989); assigned $V(OH)_2(O_2)_2^-$ by Harrison and Howarth (1985a).

vanadates and their hydrolysis products are sufficiently fast on the NMR time scale to give rise to an averaged signal (averaged with respect to the mole fractions, the chemical shifts and half-widths of the coexisting forms). The relation between the observed average chemical shift δ_{av}, the limiting shifts for HV_1 (δ_1) and H_2V_1 (δ_2), the pH and the pK_a of H_2V_1 is given by pH = pK_a + $\log(\delta_2 - \delta_{av})/(\delta_{av} - \delta_1)$ (Galeffi and Tracey, 1989).

The vanadates with the various nuclearities can be observed simultaneously in the vanadium NMR (see, e.g. Figures 4 and 6 in section 4). Attaining equilibration is in fact quite slow where vanadates V_n with n $\geqslant$ 4 are involved. Yellow decavanadate, e.g., once formed at pH < 6.5, takes several hours at room temperature to degrade to vanadates of lower nuclearities as the pH is adjusted to 7 and above. Two monovanadates and one divanadate containing peroxo groups have been detected in mixtures of vanadate and H_2O_2 in the pH range 9-5 (Campbell *et al.*, 1989). Peroxovanadates may be involved in the exchange of vanadium between the active site of bromo/iodoperoxidases from marine brown

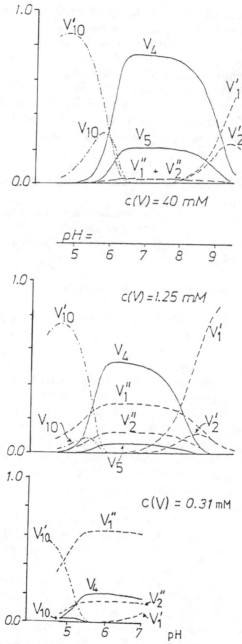

Fig. 2. Distribution diagrams, showing the pH dependence of the equilibrium concentrations of vanadates for three different overall vanadium concentrations, c(V) (mmol/L). Adapted and redrawn from Petterson *et al.* (1983 and 1985a). Dashes at the symbols indicate protonated forms; for abbreviations see Table II.

and red algae and the environment, or in the substrate (peroxide) binding by the prosthetic group of this interesting group of enzymes.

The pyro- and triphosphate analogues $HPVO_7{}^{3-}$ and $HP_2VO_{10}{}^{4-}$ are involved in rapid exchange with their hydrolysis products and give rise to broad resonance lines. The mixed anhydrides are less stable towards hydrolysis by 1 to 2 orders of magnitude than divanadate, but approximately 10^6 times more stable than pyrophosphate (Gresser et al., 1986). The relative ease of the formation of phosphovanadates is probably of some importance for the physiological role of vanadate, even though physiological concentrations of vanadate are rather low.

4. The Interaction of Vanadate with Biomimetic and Biogenic Model Compounds

Model compound studies are of considerable importance since they provide the background information for understanding the interactions of vanadate with proteins and the coordination environment of vanadium in strong vanadium-protein complexes.

4.1. CARBOXYLATES AND HYDROXYCARBOXYLATES

Structure representations of carboxylato-vanadium(V) complexes based on single crystal X-ray diffraction studies are shown in Figure 3, indicating the versatile coordination geometries and coordination modes found with carboxylates. In solution, as revealed by ^{51}V NMR, only structures **B** and **D** have been considered; so far, the coordination number 7 does not seems to have been taken into account in complex formation in the aqueous system.

Formation of complexes between vanadate and carboxylates (such as oxalate, citrate, lactate, malate, glycolate and salicylate) is strongly favored at relatively high H^+ concentrations (pH range 3-5), indicating that complex formation consumes protons. In addition, it is a general feature that the complexes once formed are further protonated as the pH decreases. These generalizations mainly apply to the octahedral complexes and not necessarily to the trigonal-bipyramidal forms for the vanadate/lactate and vanadate/glycerate systems. The complexity of the aqueous chemistry of vanadate, carboxylates, and protons is exemplified for lactate in Scheme 1 and Figure 4. Table III contains chemical shift data, assignments (some of them still tentative) of the complexes observed by NMR, and the formation constants where available.

Tetrahedral ester complexes are observed only with hydroxo acids (discussion in 4.2). The coordination numbers 4 (tetrahedral), 5 (trigonal-bipyramidal) and 6 (octahedral) for similar complexes with respect to the electronic and steric nature of the ligand can be assigned specific shifts, viz. ca. -556, ca. -517 and ca. -536, respectively. Pentavalent carboxylato-V(V) complexes with trigonal geometry have so far only been verified by X-ray analysis for $VO_2(pic)OPNMe_2$ (**D** in Figure 3); picolinato complexes will be discussed in section 4.3. Hydroxy acids can act as bifunctional chelate ligands to form dinuclear complexes (Ehde et al., 1989;

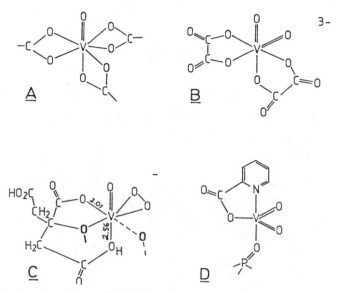

Fig. 3. Carboxylato complexes of pentavalent vanadium as revealed by single crystal X-ray structure analysis. **A**: VO(pivalate)$_3$ (Rehder *et al.*, 1989), **B**: [VO$_2$(oxalate)$_2$]$^{3-}$ (Stomberg, 1986), **C**: [{VO(O$_2$)(citrate)}$_2$]$^{2-}$ (Djordjevic *et al.*, 1989), [for clarity, only one half of the molecule plus the second bridging hydroxylate have been drawn (bridging occurs via deprotonated hydroxy groups; the dangling carboxylate and the carboxylate oxygen forming the long V-O bond are protonated)]. **D**: VO$_2$(picolinate){OP(NMe$_2$)$_3$} (Mimoun *et al.*, 1983).

Caldeira *et al.*, 1988). This has been documented by the X-ray structure of [{VO(O$_2$)μ-cit}$_2$]$^{2-}$ (Djordjevic *et al.*, 1989; **C** in Figure 3), although this complex is not directly related to divanadate (it does not contain a μ-O^{2-} functionality).

4.2. ALCOHOLS (INCLUDING SACCHARIDES AND NUCLEOSIDES), PHENOLS, AND PHOSPHATES

Simple, monovalent alcohols and phenols form mono- and diesters (with the monoesters subject to protonation equilibria), while diols such as glycol and cyclic

SCHEME 1

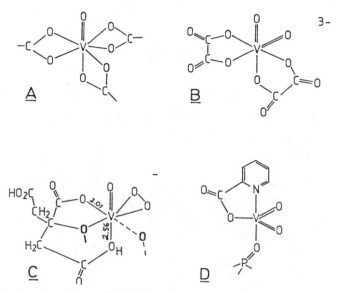

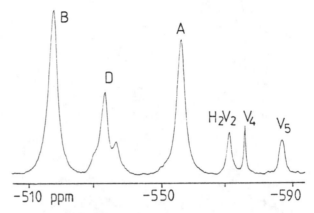

Fig. 4. 105 MHz ^{51}V NMR spectrum of 1 mM vanadate + 90 mM lactate in 8 mM Tris buffer; pH = 7.5. For abbreviations cf. Table I., for **A**, **B** and **D** Scheme 1. Redrawn from Tracey *et al.* (1987).

Table III. Data on selected vanadium(V) complexes containing carboxylato ligande.

Ligand, pH	δ	Reference	Assignment (Formation constant)
Oxalate, <7.3	− 536	Ehde *et al.* (1986)	$(H^+)_2(H_2VO_4^-)(ox^{2-})_2$
<3.8	− 533	Ehde *et al.* (1986)	$(H^+)_2(H_2VO_4^-)(ox^{2-})_2$
Oxalate, 7.04	− 536	Tracey *et al.* (1987)	$[H_2VO_2ox]^{3-}$ {13 M^{-1}} and $[VO_2(0x)_2]^{3-}$ {200 M^{-1}}
Lactate^a, 7.5	− 556	Tracey *et al.* (1987)	tetrahedral ester V_1(lac) {0.54 M^{-1}}
	− 536		octahedral V_2(lac)$_2$
	− 517		trigon. bipyr. V_1(lac)$_2$ {M^{-1}}
Lactate, 5.0	− 525	Caldeira *et al.* (1987)	(polymeric) 1 : complex
	− 544		1 : n complex (n > 1)
Malate, 4.8	− 534	Caldeira *et al.* (1987)	1 : 1 complex, mal acts as a tridentate ligand
	− 546		1 : n complex (n > 1)
Salicylate,	Galeffi and Tracey (1989)		
6.86	− 533		actahedral V_1sal {0.23 M^{-1}} and V_1(sal)$_2$ {1.7 M^{-1}}
Citrate, 8–6.5	− 547	Ehde *et al.* (1989)	$(H^+)(H_4V_2O_8^{2-})(cit^{3-})$
<6	− 543, − 551		$(H^+)(H_4V_2O_8^{2-})(cit^{3-})$
Tartrate,		Caldeira *et al.* (1988)	
7.5–3.0	− 523		trigon. bipyr. $(V_2)_2$(tar)$_2$
Pivalate^b	− 597	Rehder *et al.* (1989)	octahedral VO(piv)$_3$^{b}

^{a} See also Figure 4 and Scheme 1.
^{b} In CD$_2$Cl$_2$. The solution structure (octahedral) is not the same as the structure in the crystalline state (pentagonal bipyramidal; **A** in Scheme 1).

polyalcohols (including monosaccharides) also form trigonal-bipyramidal complexes very much reminiscent of what has been noted for the vanadate/lactate interaction (cf. Figure 4). The complexes are mononuclear (usually these are the main products) or dinuclear in vanadium. Selected chemical shift data and formation constants are listed in Table IV. Structures of special interest, discussed below, are displayed in Scheme 2. Most of the work done in this area has been

SCHEME 2

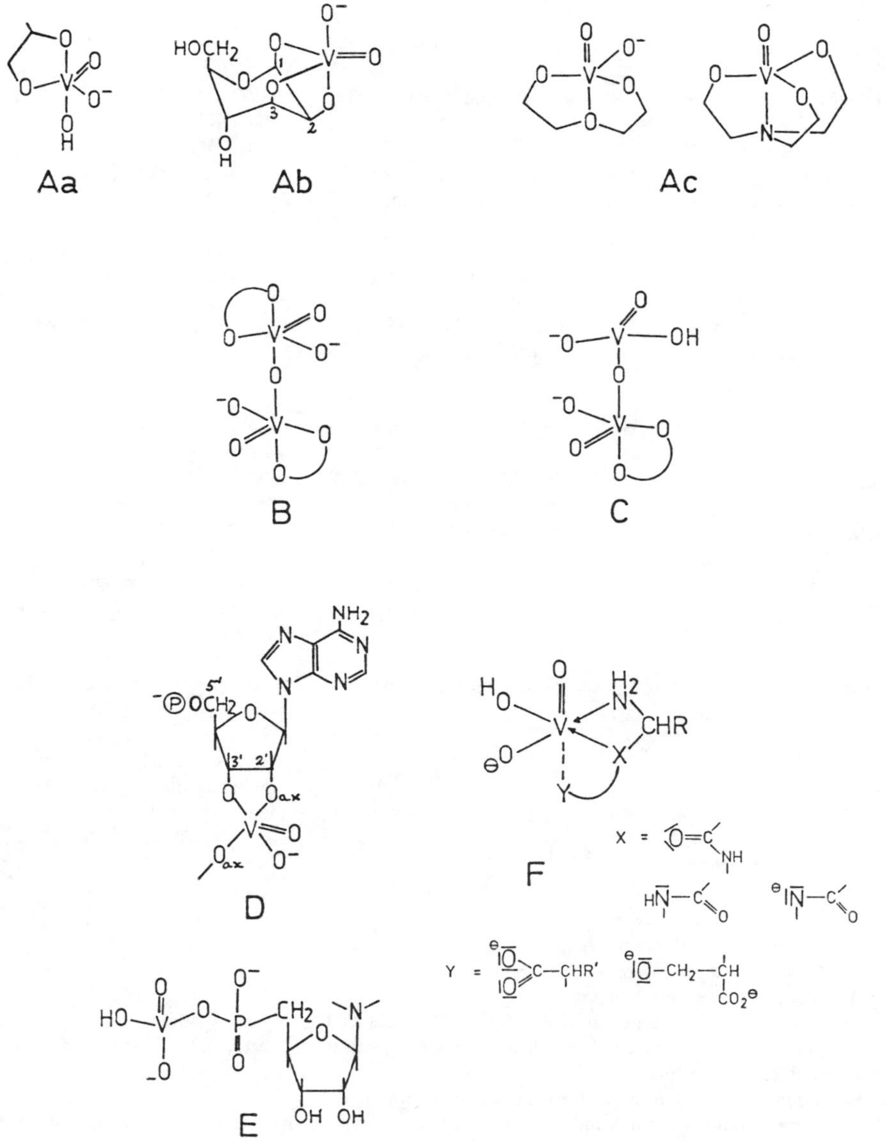

Table IV. Data for complexes formed by interactionof vanadate with hydroxy functions.

Ligand, pH[a]	$\delta(^{51}V)$	Reference	Assignment[b] (Formation constant)
MeOH	-551, -528	Tracey and Gresser (1988)	HVO_3OR^- {0.094 M^{-1}}, VO_3OR^{2-}
	-543		$VO_2(OR)_2^-$ {0.022 M^{-1}}
Me$_2$CHOH	-562, -539	Tracey and Gresser (1988)	HVO_3OR^- {0.24 M^{-1}}, VO_3OR^{2-}
	-522		$VO_2(OR)_2^-$ {0.09 M^{-1}}
Me$_3$COH	-575, -553	Tracey and Gresser (1988)	HVO_3OR^- {0.095 M^{-1}}, VO_3OR^{2-}
Phenol[c]	-565, -538	Galeffi and Tracey (1988)	HVO_3OR^- {1.8 M^{-1}}, VO_3OR^{2-}
	-571		$VO_2(OR)_2^-$ {1.1 M^{-1}}
Tyrosine[c,d]	-564, -574	Tracey and Gresser (1986)	HVO_3OR^-, $VO_2(OR)_2^-$
HOCH$_2$CH$_2$OH	-556, -533	Tracey and Gresser (1988)	HVO_3OR^- {0.34 M^{-1}}, VO_3OR^{2-}; A
	-553		$VO_2(OR_2^-$ {0.09 M^{-1}}
	-522	Gresser and Tracey (1986)	$[HVO_3(glycol)_2]\mu$-O; {$2.6 \cdot 10^5$ M^{-3} [e]}; B
O(C$_2$H$_5$OH)$_2$	-519	Crans and Shin (1988)	$VO_2(dee)^-$ {1.2 M^{-1}}; Ac
Me-β-Galactose[f]		Tracey and Gresser (1988a)	
	$-517/-521$		C
	-522		B
Mannose	$-480/-530$	Geraldes and Castro (1989)	{24[h]}
	-523[g]		Ab
Ribose	$-495/-525$	Geraldes and Castro (1989)	{230[h]}
	-517[g]		Ab
AMP[f], 7.5	-523	Tracey et. al. (1988b)	D; {$3.4 \cdot 10^6$ M^{-3}}
6.5			{$8.5 \cdot 10^6$ M^{-3}}
8.0	-551[i]		mixed anhydride APV, {5.4 M^{-1}; 19.6 at pH 6.5}; E
8.0	-569[l]		mixed anhydride APV$_2$ {1 M^{-1}; 12 M^{-1} at pH 7.5}
Uridine, 7.5	-523	Tracey et al. (1988b)	V$_2$L$_2$ (trigonal bipyramid, 2 isomers) {$2.8 \cdot 10^7$ M^{-3}}[j]
5.8	-521[k]		V$_1$L {130 M^{-1}}, V$_1$L$_2$ {123 M^{-1}}, trigonal bipyramid. involving the 2$'$ and 3$'$ OH
Guanosine, 7.5	-523, -530	Rehder et al. (1989a)	V$_1$L^1
Inosine, 7.2	-523, -509	Rehder et al. (1989a)	V$_1$L$_n^m$ {685 M^{-1}}, V$_1$L$_1^n$ {94 M^{-1}}

[a] 7.0–7.4, if not indicated.
[b] Bold letters refer to Scheme 2.
[c] In ca. 50% acetone/water.
[d] As N-acetyltyrosine ethyl ester.
[e] For the condensation reaction of two $HVO_3OCH_2CH_2OH^-$.
[f] In addition to the mixed ester-anhydrides noted here, tetrahedral esters derived from mono- and divanadate are also observed.
[g] Main component in an unresolved multi-component signal.
[h] Average formation constant under the assumption that a mononuclear 1 : 1 cyclic ester is formed.

carried out by Tracey and Gresser, and their results can be summarized, supplemented and commented, as follows:

(1) With alkyl alcohols ROH, the tetrahedral esters $ROVO_3H^-$ [-551], $(RO)_2VO_2^{2-}$ [-543], $ROV_2O_6H^{2-}$ and $(RO)_2V_2O_5^{2-}$ [-564] with the typical shift values [in brackets; quoted for R = Me and pH = 7.2] close to those of the inorganic precursors (Table II). pK_a values of the protonated esters are similar to those of the vanadates (Table II). Shift values for $ROVO_3H^-$ are pH dependent: At a pH = 10.3, a limiting value δ = -528 ppm for the deprotonated ester $ROVO_3^{3-}$ is obtained (Tracey et al., 1988). An increase in the bulk of R leads to an increase of ^{51}V shielding (Tracey and Gresser, 1988). The increase in shielding as the vanadium-ligand overlap becomes less effective as a consequence of steric crowding is a general phenomenon observed with V(V) (see also section 2.), and has been investigated systematically for the complexes $VO(oxine)_2OR$ (Weidemann et al., 1988).

(2) Constants of formation, K, for the esters can be obtained by quantitatively evaluating the concentration dependencies of the relative integral signal intensities of the vanadium containing species in equilibrium with each other (an example for this evaluation will be given in section 4.4.). For vanadate esters formed with simple alkyl alcohols, K values are typically 0.1-0.3 M^{-1} for the mononuclear monoester and 0.02-0.1 for the diesters (Tracey and Gresser, 1988). K is related to the pK_a of the alcohol (larger formation constant for the higher pK_a) (Tracey et al., 1988a). Phosphate catalytically enhances hydrolysis of the esters, possibly via the formation of mixed vanadate-phosphate esters (Tracey et al., 1988).

(3) It is known that vanadate esters are accepted as substrates by some enzymes. More specifically, vanadate(V), which has been ascribed an insulin-like effect, activates the phosphorylation of a tyrosine residue in the insulin receptor, and this may well be the result of an esterification of the tyrosine hydroxyl by vanadate (Gresser et al., 1987). In this context, the investigation of aryl esters of vanadate is of special interest (Tracey and Gresser, 1986; Galeffi and Tracey, 1988). Phenol and phenol derivatives (including tyrosine) form vanadate esters analogous to those observed with alkyl alcohols. The formation constants are, however, about five times as large, and ^{51}V shielding exceeds that of alkyl esters by ca. 20 ppm (see Table IV for data).

(4) A more complex product pattern is observed with alcohols containing two and more hydroxyl groups (Gresser and Tracey, 1986; Tracey and Gresser, 1988a; Tracey et al., 1988b; Geraldes and Castro, 1989). Again, esters com-

---

[i] Superimposed with the signals for vanadate and vanadate esters in rapid exchange with the mixed AMP-vanadate anhydrides.

[j] For the formation from 2 monovanadates and two uridine ligands.

[k] $W_{1/2}$ = 580 Hz.

[l] The signal at -530 is attributed to the coordination through the NH_2 group of guanine.

[m] For the concentration ratios vanadate/inosine <1.5, n = 1. Line width at half-height, $W_{1/2}$ = 910–1050 Hz.

[r] $W_{1/2}$ = 500 Hz.

parable to those obtained with monoalcohols are present, with essentially the same chemical shifts and formation constants (Table IV). In addition, cyclic structures are observed, arising from the formation of the thermodynamically favored chelate-5 rings. The geometry of these cyclic esters, derived from mono- (Scheme 2, **Aa**) and divanadate (**B** and **C**), presumably is trigonal-bipyramidal, with δ values around -522 ppm. In the case of D-Mannose, it has been shown by ^{1}H NMR that a cyclic triester is formed. The OH groups of C1, C2 and C3 of β-mannopyranoside in its distorted ^{4}C1 boat conformation form ester linkages to vanadate in the main species represented by $\delta(^{51}V)$ = -523 ppm, and attributed a trigonal-bipyramidal structure (**Ab** in Scheme 2) (Geraldes and Castro, 1989). Mixed tetrahedral/trigonal-bipyramidal complexes like the one shown in **C**, Scheme 2, have been deduced from the evaluation of the ^{51}V NMR spectra in the case of methyl-β-galactopyranoside (Tracey and Gresser, 1988a). Primary and secondary alcoholic functions of a saccharide are involved in the tetrahedral and trigonal-bipyramidal species.

In octahedral complexes with five of the vanadium coordination sites blocked by appropriate ligands, the preferred mode for esterification for a saccharide such as ribose is via the dangling, i.e. primary hydroxyl (Weidemann et al., 1988). An example is shown in Figure 5 for the ester formed between the (hypothetic) acid $VO(oxine)_2OH$ and desoxyuridine.

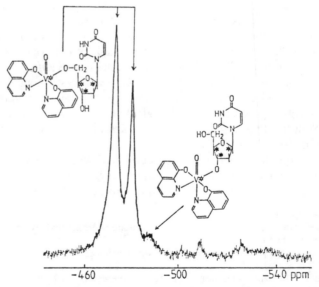

Fig. 5. 94.7 MHz ^{51}V NMR spectrum of the esters $VO(oxine)_2(OR)$ with HOR = desoxyuridine in CD_2Cl_2. Desoxyuridine coordinates almost exclusively through the primary alcoholic function, giving rise to two signals belonging to two diastereomeric pairs of enantiomers. An asterisk indicates a center of chirality. From Weidemann et al. (1988).

The formation of vanadate esters with penta-coordinated vanadium in a trigonal-bipyramidal arrangement is of considerable importance in the light of the transition state postulated for phosphorus during enzymatic hydrolysis of the phospho-diester bond in RNA (see also section 5), and the investigations with polyalcohols have therefore been extended to nucleosides and nucleotides such as adenosine, cytidine, uridine, guanosine, inosine and adenosinephosphates (Tracey et al., 1988b; Geraldes and Castro, 1989, 1989a; Gil, 1989; Rehder et al., 1989a). The predominant ester species formed between vanadate and AMP is a dimeric compound containing two vanadium ions and two AMP ligands. The latter are coordinated through the 2'- and 3'-hydroxyls of the ribose ring (D in Scheme 2). Competing with the esterification reaction, and favored by a factor of about 40, is the formation of mixed anhydrides, leading to the ADP and ATP analogs AVP (E in Scheme 2) and AV_2P (formation constants at pH $= 7.5$ are 8.8 and 12 M^{-1}, respectively; the formation constants drastically decrease with increasing pH). Owing to rapid formation/degradation, the signals for the mixed anhydrides are rather broad. A similar effect has been discussed in the context of phospho-vanadates (section 3.).

The nitrogen functions of the adenine moiety are not involved in coordination to vanadium, and similar observations have been reported for uridine (Tracey and Gresser, 1988b) and inosine (Rehder et al., 1989a). Reaction with the 2'- and 3'-hydroxyls of the ribose ring apparently is "an optimum geometrical relationship for condensation to occur" (Tracey and Gresser, 1988b): The constant for the formation of the trigonal-bipyramidal (dinuclear and biligate) compound is 9 orders of magnitude larger than the analogous value determined for the glycol complex. Guanosine, which differs from the other nucleosides by carrying an amino substituent in the 2-position, i.e. adjacent to a secondary ring nitrogen, gives rise to a resonance in the ^{51}V NMR which probably corresponds to a complex with N-coordination (Rehder et al., 1989a).

4.3. AMINES AND PYRIDINES

Data of selected complexes are collated in Table V. Pyridine gives rise to a weak signal at -467 ppm, ascribed to a tetrahedral complex containing two pyridines (Galeffi and Tracey, 1989), the shift apparently reflecting coordination of nitrogen. Two octahedral complexes, mononuclear in vanadium, are observed with 2-hydroxy nicotinic acid, which coordinates through the hydroxy and carboxylate oxygens in the same manner as lactate (Figure 4) and salicylate. Comparable complexes are obtained with 3-hydroxy picolinate, but in addition, carboxylate + nitrogen coordination is observed as with picolinate itself. The formation of the favored chelate-ring structures gives rise to complexes which are considerably more stable than those formed with (monodentate) carboxylates, alcohols or pyridine. The 1 : 1 complex formed between picolinate and monovanadate at pH $= 7$ presumably has the trigonal bipyramidal geometry verified for the picolinato complex D in Scheme 1 by an X-ray structure analysis. The formation of these complexes is strongly pH dependent and cannot be observed at pH > 8.

Table V. Data for complexes formed by interaction of vanadate with nitrogen containing ligands (see also text for additional data).

Ligand, pH[a]	$\delta(^{51}V)$	Ref.	Assignment (Formation constant)
2-OH-nicotinate	-535	b	oct. V_1L $\{0.58\ M^{-1}\} + V_1L_2$ $\{17\ M^{-1}\}$
3-OH-nicotinate	-536	b	oct. V_1L $\{570\ M^{-1}\} + V_1L_2$ $\{44\ M^{-1}\}$
			(coordination via $-O^-$ and $-CO_2^-$)
	$-549, -512$		V_1L $\{110\ M^{-1}\}$, V_1L_2 $\{690\ M^{-1}\}$
			(coordination via N and CO_2^-)
Picolinate	-550	b	V_L $\{200\ M^{-1}\}$ (trigonal bipyramid?)
	$-552, -529, -513$		V_1L_2 $\{1.5\text{-}, 4.0\text{-}, 8.4 \cdot 10^4\ M^{-2}\}$
Diethanolamine	-488	c	V_1L $\{510\ M^{-1}\}$
Triethanolamine	-483	c	V_1L $\{2000\ M^{-1}\}$
Iminodiacetate	-516	c	V_1L
N-OH-ethyl-glycine	-506	c	V_1L
Tris[e], 7.6	-500	d	$V_1(Tris)_2$ $\{2.6\ M^{-2}\}$
	-534		trigonal-bipyramidal $V_2(Tris)_2$

[a] See footnotes a in Table IV.
[b] Galeffi and Tracey (1989).
[c] Crans and Shin (1989).
[d] Tracey and Gresser (1988b).
[e] c(Vanadate) = 2 mM, c(Tris) = 400 mM, ionic strength = 1 M (maintained with KCl). Other complexes than those indicated are also present.

The products of interaction of vanadate with a large variety of potentially tri- and tetradentate amine ligands have been investigated, containing, in addition to the amine nitrogen, functional groups such as OH, CO_2H, PO_3H^- or NH_2 (Crans and Shin, 1988). For molar ratios vanadate/ligand up to 1/10, only one type of complex (mononuclear and monoligate, δ = -480 to -516 ppm) is formed. An example is displayed in **Ac**, Scheme 2. The chemical shifts are to low field of the vanadate complexes discussed so far, possibly as a consequence of the more easily polarizable nitrogen functionality (cf. section 2.). Formation constants are 2 to 3 orders of magnitude greater than for complexes with diols or ether alcohols (Table IV) and hydroxy acids (Table III), provided that a lone pair on the bridge-head nitrogen is available, and at least two of the additional functionalities are terminal groups of an ethylene backbone. This considerably increased thermodynamic stability of the complexes has been attributed to the coordination of the amine nitrogen.

Many of these amino polyalcohols are in use as, or related to buffer systems in biochemistry. A typical example is tris(hydroxymethyl)aminomethane (Tris) (Tracey and Gresser, 1988b). Although complex formation on a whole is weak (neither the chelate-4 ring formed as Tris coordinates via the amino and hydroxyl groups, nor the chelate-6 rings to be expected for the coordination as a diol are thermodynamically favored), several products can be rationalized on the basis of an evaluation of the ^{51}V NMR pattern. At pH $\geqslant$ 7.6, tetrahedral, octahedral and trigonal-bipyramidal complexes are observed, both mono- and dinuclear, and mono- and biligate. At pH = 7.6, a vanadate concentration of 2 mmol/L, a 200-fold molar excess of Tris, and an ionic strength corresponding to 1 M KCl, there are two resolved signals at δ = -500 and -534 ppm. The resonance of the latter, comprising a binuclear, biligate complex (for the proposed structure cf. **B** in Scheme 2), is strongly pH-dependent (pH = 8.8: -527, pH = 6.4: -555 ppm). Two broad and badly resolved signals (at -515 and -555 ppm) appear at pH < 7.6. For 0.2 M Tris/HCl at pH = 7.2, containing 10 mmol/L of vanadate, we have observed two resonances belonging to vanadate-Tris complexes, viz. -528 ppm ($W_{1/2}$ = 350 Hz) and -544 ppm (very broad), in the approximate intensity ratio 0.6/1 (Rehder et al., 1989a).

4.4. DI- AND TRIPEPTIDES

Amino acids interact weakly only, and no systematic investigations have so far been carried out. Applied in large excess, histidine, at pH = 7.6, gives rise to a resonance of a weak vanadate-His complex at -545 ppm (Rehder et al., 1989a). Serine, again in large excess, reacts as a hydroxycarboxylic acid (Rehder, 1988). On the other hand, some oligopeptides containing His, such as Gly-His, have a rather high affinity to vanadate(V), comparable to that of the amino alcohol derivatives discussed above, while others, like His-Gly, do not form vanadate complexes at all (Rehder et al., 1989a). A collection of shift data, half-widths and complex formation constants is given in Table VI.

Since most amino acids themselves do not readily interact with vanadate, the peptide linkage appears to have a crucial role for complex formation, and there are arguments for an involvement of the amide nitrogen (instead of the carbonyl

Table VI. Data on vanadate-peptide complexes.

Ligand, pH	$\delta(^{51}V)$ (ppm)	$W_{1/2}$ (Hz)	Ref.	Formation constant (M^{-1})
Phe-Glu, 7.5	− 500		b	270 [a]
Val-Glu, 7.5	− 511	670	c	89
Gly-Asp, 7.5	− 509	410	d	49
Gly-Gly, 7.5	− 504	360	d	16
Gly-Tyr, 6.8	− 510	780	d	
7.2	− 510			142
8.6	− 507, − 500			
Gly-Ser, 7.5	− 493, − 506	420	d	
Gly-His, 7.3	− 511	520	c	113
Gly-His-Lys, 7.3	− 528		c	11
Pro-Gly, 7.4	− 493	450	d	90
Pro-His-Ala, 7.3	− 518	550	c	36

[a] For molar ratios Phe-Glu/vanadate < 3.5.
[b] Rehder et al. (1989b).
[c] Rehder et al. (1989a).
[d] Rehder (1988).

oxygen) of the peptide functionality: With tertiary amide functions (sarcosine, Gly-Pro), no resonances owing to vanadate-peptide complexes have been actually observed (while Pro-Gly forms a relatively strong complex, Table VI). Further, if *N*-protected peptides are employed, interaction between vanadate and the dipeptide is again very weak, and this is also true, if the neighborhood of a potential coordinating function is sterically shielded (Val-Gly). These findings suggest participation of the terminal NH_2. The terminal carboxylate does not seem to be substantially involved, but side-chain functions might, if they are available from a sterical and statistical point of view. Examples are Gly-Ser and Val-Glu (but not Val-Asp, with the shorter *C*-chain for the dangling carboxylate). Based on these considerations deduced from [51]V NMR studies, the coordination numbers 5 to 6 and the coordination modes as displayed in **F**, Scheme 2, have been proposed (Rehder, 1988; Rehder *et al.*, 1989b).

The stoichiometry of complex formation can be derived from a concentration and pH dependent study as shown for Gly-Tyr in Figures 6-8: Evaluation of the relative integral intensities of the resonance signals (Figure 6) and their graphical

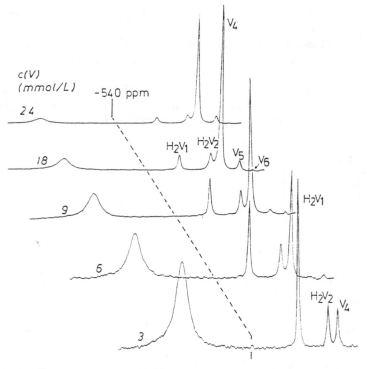

Fig. 6. 97.4 MHz [51]V NMR spectra of solutions containing 30 mmol/L of Gly-Tyr and varying amounts of vanadate [c(V), indicated on the left hand margin]. The pH has been adjusted to 7.2. The broad signal at ca. -510 ppm belongs to a vanadate-peptide complex, which is the dominant species at high peptide/vanadate ratios. For the sharp signals, which represent uncoordinated, tetrahedral vanadates, cf. Table II. (From Rehder *et al.*, 1989b).

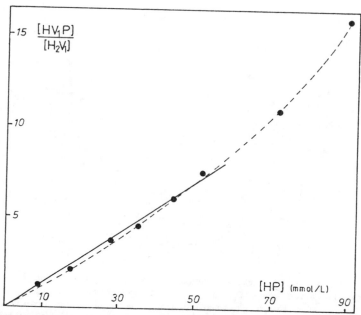

Fig. 7. Determination of the complex formation constant at pH 7.2 for the formation of the vanadate/Gly-Tyr complex, HV_1P, from H_2V_1 and HP, defined by $K = [HV_1P]/[H_2V_1][HP]$ (straight, solid line) for the equilibrium $H_2V_1 + HP \rightleftharpoons HV_1P + H_2O$ The equilibrium concentrations have been obtained by relating the relative integral intensities of the resonances such as shown in Fig. 6 to the initial concentrations [c(V) = 10 mM, c(HP) = 10-100 mM]. The increasingly noticable curvature for [HP] > 50 mmol/L indicates that, with larger peptide/vanadate ratios, the complex formation becomes more complex.

presentation (Figure 7) allows for the evaluation of the nuclearity of the complex, the stoichiometry of the reaction, and the formation constant. Complex stability usually decreases with increasing pH. In some cases, new species arise, as with glycyl-tyrosinate (Figure 8), where the phenolate-oxygen becomes increasingly available with increasing pH. It readily coordinates to vanadate (Rehder *et al.*, 1989b) in a manner, however, clearly different from what has been observed with phenols and *N*-acetyltyrosin ester (Tracey and Gresser, 1986; Galeffi and Tracey, 1988; Table IV). While, for the latter, $\delta(^{51}V)$ values of ca. -574 have been reported and allocated to tetrahedral vanadate esters, the resonance of the vanadate/ Gly-Tyr complex arising at higher pH is at -500 ppm and belongs to a complex as shown in F, Scheme 2.

5. Interaction of Vanadate with Biogenic Molecules *in vitro*

With one exception, only interactions with proteins have been investigated by the use of ⁵¹V NMR: Vanadate binding to bovine serum albumin (Vilter and Rehder,

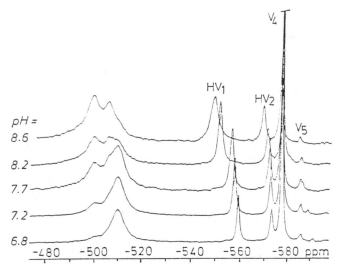

Fig. 8. 97.4 MHz ^{51}V NMR spectra of solutions of varying pH (left hand margin), containing constant concentrations of vanadate (10 mM) and Gly-Tyr (40 mM). The low-field signal arising at higher pH has been allocated a complex where the phenolic functionality participates in coordination. Also subject to pH changes are the resonances for mono- and divanadate (cf. Table II). (From Rehder *et al.*, 1989b).

1987) and apotransferrin (Butler and Eckert, 1989), vanadate-dependent halo-peroxidases (Vilter and Rehder, 1987; Rehder *et al.*, 1987), and vanadate inter-action with phosphoglycerate mutase (Stankiewicz *et al.*, 1987), alkaline phos-phatase (Crans *et al.*, 1989) and ribonucleases (Rehder *et al.*, 1989a; Borah *et al.*, 1985). The exception is a study of vanadate binding to desferrioxamine B (Butler *et al.*, 1989). This siderophore contains 3 hydroxamic acid residues and two amide functions. An ATPase assay (pH = 7.4) containing desferrioxamine and vanadate shows a ^{51}V NMR signal at -515 ppm ($W_{1/2}$ = 770 Hz) belonging to a *strong* 1:1 complex. The δ^{51}V value is in the chemical shift range commonly observed for complexes originating from amino carboxylic acids (cf. Tables V and VI), and for trigonal-bipyramidal complexes formed with *OO*-functional ligands such as lactic acid and diols (Tables III and IV).

Butler and Eckert (1989) have shown that ^{51}V NMR is a feasible probe to study metal binding to human serum transferrin (TF), where the coordination environ-ment involves the oxygens of tyrosine and aspartate, and histidine-*N*. There are two slightly inequivalent binding sites for a metal ion in transferrin, located at the *C*- and the *N*-terminal end, M_C-TF and M_N-TF, respectively. Up to 1.7 vanadates are consumed per molecule apotransferrin, and two overlapping but distinct signals at -529.5 (V_C-TF) and -531.5 (V_N-TF) with an overall widths of 420 Hz. The assignments have been carried out on the basis of the signals observed as vanadate is reacted with Fe$_N$-TF and Fe$_C$-TF. The resonance position for V_N-TF is slightly temperature dependent, and both signals shift to high field by about 5 ppm as the external magnetic field decreases from 11.7 to 7.05 T ("second order frequency shift"). The signal positions are in accord with octahedral vanadium coordination as provided by histidine residues (Table VI, section 4.4.) and ligands containing carboxylate plus phenolic functions such as salicylate (Table III, section 4.1.).

The narrow resonance lines for vanadium embedded in a huge protein molecule indicate that the system, as a consequence of the long molecular correlation time τ_c of TF ($2 \cdot 10^{-7}$ s), is not in the extreme narrowing regime. In this case, the spin-spin relaxation time T_2 is no longer governed by the quadrupole relaxation mechanism (eq. 1, section 2) but, depending on the strength of the magnetic field B_o, by chemical shift anisotropy (high B_o) or second order quadrupole interaction, and the central transitions ($+1/2 \rightarrow -1/2$) becomes detectable due to a favorable relaxation time in the ms range (line widths of several hundred Hz). The validity of this assumption has been testified by looking at the dependence of $W_{1/2}$ upon (i) the viscosity of the solution (no effect observed) and (ii) the magnetic field strength: The observed increase of $W_{1/2}$ with decreasing field (Butler and Eckert, 1989) indicates that in V-TF second order quadrupole perturbation is the predominant line broadening factor.

By contrast, rather broad ^{51}V resonances (ca. 5 kHz) arise for vanadate(V) bound to bovine serum albumine [δ = -710 to -970 ppm, depending on the vanadate/albumine ratio (Vilter and Rehder, 1987)] and in vanadate-dependent bromo/iodoperoxidase from the brown sea-weed Ascophyllum nodosum, An, (Rehder and Vilter, 1987; Rehder et al., 1987), for which the temperature dependence of $W_{1/2}$ shows that quadrupole relaxation dominates and hence eq. 1 applies. Since An and TF have similar molecular masses (ca. 100,000 g/mol), the size of the molecule cannot be soly responsible for the different appearances of the ^{51}V resonances. Another striking difference is the very high shielding of the ^{51}V nucleus in V-An, varying between δ = -1100 and -1300, with the greatest shielding observed in the presence of an excess of the substrate iodide. The high shielding has led to the assumption, based on the grounds considered in section 2, that the coordination sphere of vanadium in V-An is occupied by oxygen atoms, with two or three strained structure elements providing an additional shielding effect. Chelate-4 rings, such as formed with bidentate carboxylato groups of acidic side chains in the protein, might be responsible for strained structures. The complex oxo-tris(pivalato)vanadium shown in A, Figure 3 (section 4.1.) is the best model presently available for the active site in V-An, although, in solution, the ^{51}V nucleus is only slightly more shielded [$\delta(^{51}V)$ = -597 ppm (Rehder et al., 1989)] than in other oxovanadium complexes. This is, however, a consequence of the instability of the 4-ring structures in solution which, in an overall equilibrium encompassing several species, contribute to the shielding by a few percent only.

While vanadyl is a strongly bond constituent in V-An, this is not neccessarily so for the binding of vanadate to phosphorylating enzymes. Vanadate concentrations leading to a change in activity of enzymes such as alkaline phosphatase, phosphoglycerate mutase or ribonucleases are in the range 10^{-5} to 10^{-7} M, and under these conditions, vanadium is exclusively present as monovanadate (compare Figure 2 in section 3). ^{51}V NMR can be used - and has been used - to determine the concentration of monovanadate in enzyme kinetics, e.g. to verify the correlation between the rate of hydrolysis of p-nitrophenylphosphate catalyzed by alkaline phosphatase and the concentration of the inhibiting vanadate (Crans et al., 1989).

It has been known for some time that 2,3-diphosphoglycerate phosphatase activity of phosphoglycerate mutase (PGM) is stimulated by vanadate. The activity is further enhanced by a factor of 3 as vanadate and phosphate are added. This effect is probably a consequence of the formation of a mixed phosphate-vanadate anhydride (phosphovanadate; see section 3). Concomitantly, an increase of the oxygen affinity of human red blood cells is observed, a fact which has stimulated expectations as to the possible use of vanadate as a therapeutic agent in the treatment of sickel cell anemia. A study of vanadate binding to PGM by ^{51}V NMR (Stankiewicz et al., 1987) has revealed a binding capacity of 4 vanadium ions per molecule of PGM, and from the set of data it has been rationalized that, in the concentration range considered [c(V) = 0.1-1.5 mM, c(PGM) = 0.15 mM, pH = 7.0]. Two divanadates bind non-cooperatively to each of the two subunits of PGM, corresponding to $PGM(V_2)_2$ with an intrinsic formation constant of $4 \cdot 10^6$ M^{-1}. The structure proposed for the coordination of divanadate to the active site of PGM, displayed in Scheme 3, **A**, takes into account the fact that approximately one half of the enzyme-bound vanadate is NMR-visible (broad resonance in the region of oligo-vanadates, corresponding to the "dangling", tetrahedral moiety), while the other half, directly attached to the protein through a nitrogen function, is NMR-invisible owing to quadrupole broadening beyond detection.

The pentagonal-bipyramidal geometry for the directly bonded vanadium site in $PGM(V_2)$ has also been rendered plausible from the results of a neutron diffraction study for the ternary complex formed between bovine pancreatic ribonuclease A (RNase-A), uridine and vanadate (**B** in Scheme 3). This suggestion also conforms with the chemical shift $\delta(^{51}V)$ = -522 ppm found for this complex (Borah et al., 1985). Since the same chemical shift is observed for the binary vanadate-uridine complex [a cyclic ester with vanadate binding through the oxygens at the 2' and 3' positions of the ribose ring; cf. section 4.2.(4)], this feature is considered to be maintained on addition of RNase-A to the binary compound.

Studies of this kind are being carried out to mimick the postulated transition state for the hydrolysis of the phospho-diester bond in ribonucleic acid. Despite the chemical similarity between phosphate and vanadate, vanadium, as a transition metal, exhibits a much stronger tendency to stabilize the coordination number 5. A similar situation as that reported for RNase-A is encountered with the interaction between vanadate, ribonuclease T_1 (RNase-T_1) from Aspergillus oryzae, and guanosine or inosine. RNase-T_1 specifically cleaves RNA at the 3'-end of guanosine and is inhibited by vanadate. The complex formation between RNase-T_1 and $H_2VO_4{}^-$ has been detected by ^{51}V NMR (Figure 9, bottom trace; Rehder et al., 1989a). Average formation constant [145(30) M^{-1}] and chemical shift (-514 ppm) are similar to those of the complexes formed between mono-

SCHEME 3

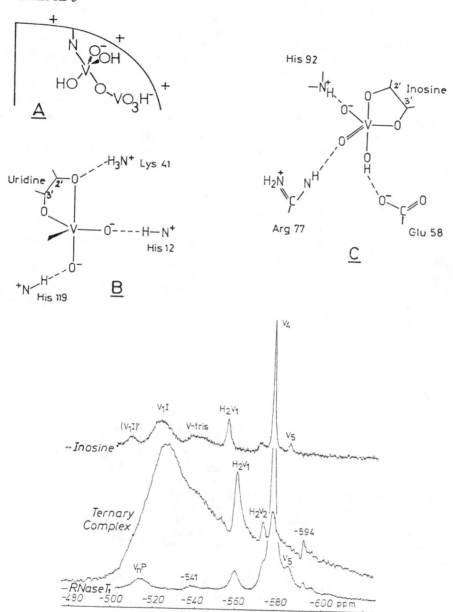

Fig. 9. 94.7 MHz ^{51}V NMR spectra of the systems (from top to bottom) vanadate/inosine (pH 7.8), vanadate/inosine/RNase-T$_1$ (pH 7.2) and vanadate/RNase-T$_1$ (pH 7.2) in Tris/HCl buffer. Approximate concentrations are c(V) = 12, c(inosine) = 11, c(RNase-T$_1$) = 6 mM. The signals labeled V$_n$P and V$_1$I belong to the vanadate complexes; for the free species see Table II. From Rehder *et al.* (1989a).

vanadate and small peptides (section 4.4.). This observation suggests a similar complexation behavior in solution. In crystals of the binary complex RNase-T$_1$/ vanadate, tetrahedral $H_2VO_4^-$ has been detected at the active site, bonded to various active site functions through hydrogen bonds (Kostrewa *et al.*, 1989). Addition of inosine led to the formation of the ternary complex, with the resonance line considerably broadened and shifted to -523 ppm (middle trace in Figure 9). The spectral parameters now are similar to those of the binary complex formed between vanadate and guanosine or inosine (upper trace), leading to the assumption that vanadate is directly bonded to the nucleotide in the usual manner (C in Scheme 3), while additional interaction through hydrogen bonds to the amino acid residues at the active site of RNase-T$_1$ tightly incorporates the enzyme in the ternary compound. The strong bonding interaction is reflected in the formation constant ($1.5 \cdot 10^5$ M^{-2}) and also evident from the fact that the overall concentration of free, "inorganic" vanadate decreases to a point where *mono*vanadate is the main component (whereas in the binary complexes the species is tetravanadate; cf. Figure 9).

In addition to the above mentioned protein studies, ^{51}V NMR has also been used to probe the in vivo metabolism of vanadium in cells of the yeast *S. cerevisiae* (Willsky *et al.*, 1984, 1985; Willsky and Dosch, 1986). These studies have helped to identify the vanadate species responsible for the cellular toxicity of vanadium. The intracellular reduction of vanadate(V) to vanadyl(IV) appears to be a mechanism of detoxification.

References

Borah, B.; Chen, C. W.; Egan, W.; Millar, M.; Wlodawer, A.; Cohen, J. S.: 1985, *Biochem.* **24**, 2058-2067.

Butler, A.; Eckert, H.: 1989, *J. Am. Chem. Soc.* **111**, 2802-2809.

Butler, A.; Parsons, S. M.; Yamagata, S. K.; de la Rosa, R. I.: 1989, *Inorg. Chim. Acta* **163**, 1-3.

Caldeira, M. M.; Ramos, M. L.; Oliveira, N. C.; Gil, V. M. S.: 1987, *Can. J. Chem.* **65**, 2434-2440.

Caldeira, M. M.; Ramos, M. L; Oliveira, N. C.; Gil, V. M. S.: 1988, *J. Molec. Struct.* **174**, 461-466.

Campbell, N. J.; Dengel, A. C.; Griffith, W. P.: 1989, *Polyhedron* **11**, 1379-1386.

Crans, D. C.; Bunch, R. L.; Theisen, L. A.: 1989, *J. Am. Chem. Soc.* **111**, 7597-7607.

Crans, D. C.; Shin, P. K.: 1988, *Inorg. Chem.* **27**, 1797-1806.

Djordjevic, C.; Lee, M.; Sinn, E.: 1989, *Inorg. Chem.* **28**, 719- 723.

Ehde, P. M.; Andersson, I.; Pettersson, L.: 1986, *Acta Chem. Scand.* **A40**, 489.

Ehde, P. M.; Andersson, I.; Pettersson, L.: 1989, *Acta Chem. Scand.* **43**, 136-143.

Galeffi, B.; Tracey, A. S.: 1988, *Can. J. Chem.* **66**, 2565-2569.

Galeffi, B.; Tracey, A. S.: 1989, *Inorg. Chem.* **28**, 1726-1734.

Geraldes, C. F. G. C.; Castro, M. M. C. A.: 1989, *J. Inorg. Biochem.* **35**, 79-93.

Geraldes, C. F. G. C; Castro, M. M. C. A.: 1989a, *J. Inorg. Biochem.*, in press.

Gil, V. M. S.: 1989, *Pure Appl. Chem.* **61**, 841-848.

Gresser, M. J.; Tracey, A. S.: 1986, *J. Am. Chem. Soc.* **108**, 1935-1939.

Gresser, M. J.; Tracey, A. S.; Parkinson, K. M.: 1986, *J. Am. Chem. Soc.* **108**, 6229-6234.

Gresser, M. J.; Tracey, A. S.; Stankiewicz, P. J.: 1987, *Adv. Prot. Phosphatases* **4**, 35-57.

Harrison, A. T.; Howarth, O. W.: 1985, *J. Chem. Soc. Dalton Trans.*, 1953-1957.

Harrison, A. T.; Howarth, O. W.: 1985a, *J. Chem. Soc., Dalton Trans.*, 1173.

Howarth, O. W.; Pettersson, L.; Andersson, I.: 1989, *J. Chem. Soc., Dalton Trans.*, in press.

Ivakin, A. A.; Kurbatova, L. D.; Kruchinina, M. V.; Medvedeva, N. I.: 1986, *Russ. J. Inorg. Chem.* **31**, 219-222.

Kostrewa, D.; Choe, H.-W.; Heinemann, U.; Saenger, W.: 1989, *Biochemistry* **28**, 7592-7600.

Maksimovskaya, R. I.; Chumachenko, N. N.: 1987, *Polyhedron* **6**,1813-1821.

Mimoun, H.; Saussine, L.; Daire, E.; Postel, M.; Fischer, J.; Weiss, R.: 1983, *J. Am. Chem. Soc.* **105**, 3101-3110.

Paulsen, K.; Rehder, D.: 1982, *Z. Naturforsch.* **37a**, 139-149.

Pettersson, L.; Andersson, I.; Hedman, B.: 1985, *Chem. Scr.* **25**, 309-317.

Pettersson, L.; Hedman, B.; Andersson, I.; Ingri, N.: 1983, *Chem. Scr.* **22**, 254-264.

Pettersson, L.; Hedman, B.; Nenner, A.-M.; Andersson, I.: 1985, *Acta Chem. Scand.* **A39**, 499-560.

Priebsch, W.; Rehder, D.: 1985, *Inorg. Chem.* **24**, 3058-3062.

Rehder, D.: 1982, *Bull. Magn. Reson.* **4**, 33-83.

Rehder, D.: 1984, *Magn. Reson. Rev.* **9**, 125-237.

Rehder, D.: 1987, in J. Mason (ed.) *Multinuclear NMR*, Plenum Press, New York, ch. 19.

Rehder, D.: 1988, *Inorg. Chem.* **27**, 4312-4316.

Rehder, D.; Priebsch, W.; von Oeynhausen, M.: 1989, *Angew. Chem.* **101**, 1295-1297.

Rehder, D.; Holst, H.; Quaas, R.; Hinrichs, W.; Hahn, U.; Saenger, W.: 1989a, *J. Inorg. Biochem.*, in press.

Rehder, D.; Weidemann, C.; Priebsch, W.; Holst, H.: 1989b, *27th Int. Conf. Coord. Chem.*, Gold Coast (Australia); see *Conf. Abstr.*, p. M60.

Rehder, D.; Weidemann, C.; Duch, A.; Priebsch, W.: 1988, *Inorg. Chem.* **27**, 584-587.

Rehder, D.; Vilter, H.; Duch, A.; Priebsch, W.; Weidemann, C.: 1987, *Rec. Trav. Chim. Pays-Bas* **106**, 408.

Stankiewicz, P. J.; Gresser, M. J.; Tracey, A. S.; Hass, L. F.: 1987, *Biochem.* **26**, 1264-1269.

Stomberg, R.: 1986, *Acta Chem. Scand.* **A40**, 168.

Tracey, A. S.; Gresser, M. J.: 1986, *Proc. Natl. Acad. Sci. USA* **83**, 609-613.

Tracey, A. S.; Gresser, M. J.: 1988, *Can. J. Chem.* **66**, 2570-2574.

Tracey, A. S.; Gresser, M. J.: 1988a, *Inorg. Chem.* **27**, 2695-2702.

Tracey, A. S.; Gresser, M. J.: 1988b, *Inorg. Chem.* **27**, 1269-1275.

Tracey, A. S.; Gresser, M. J.; Galeffi, B.: 1988, *Inorg. Chem.* **27**, 157-161.

Tracey, A. S.; Galeffi, B.; Mahjour, S.: 1988a, *Can. J. Chem.* **66**, 2294-2298.

Tracey, A. S.; Gresser, M. J.; Liu, S.: 1988b, *J. Am. Chem. Soc.* **110**, 5869-5874.

Tracey, A. S.; Gresser, M. J.; Parkinson, K. M.: 1987, *Inorg. Chem.* **26**, 629-638.

Tracey, A. S.; Radley, K.: 1985, *Can. J. Chem.*, **63**,2181-2184.

Unkel, P.; Buch, P.; Dembczynski, J.; Ertmer, W.; Johann, U.: 1989, *Z. Phys. D.: At. Mol. Clusters* **11**, 259-271.

Vilter, H.; Rehder, D.: 1987, *Inorg. Chim. Acta* **136**, L7-L10.

Weidemann, C.; Priebsch, W.; Rehder, D.: 1988, *Chem. Ber.* **122**, 235-243.

Willsky, G.R.; Dosch, S.F.: 1986, Yeast **2**, 77-85.

Willsky, G.R.; Leung, J.O.; Offermann, Jr., P.V.; Plotnick, E.K.: 1985, J. Bacteriol. **164**, 611-617.

Willsky, G.R.; White, D.A.; McCabe, B.C.: 1984, J. Biol. Chem. **259**, 13273-13281.

XI. Biological Applications of EPR, ENDOR, and ESEEM Spectroscopy

SANDRA S. EATON and GARETH R. EATON
Department of Chemistry, University of Denver, Denver, CO 80208, U.S.A.

1. Introduction

Electron paramagnetic resonance (EPR) has the special feature that it selectively detects only paramagnetic species, and reveals information about the immediate environment and dynamics of that paramagnetic species, even when it is part of a much larger complex species. Hence EPR powerfully ignores, for example, the bulk of a metalloprotein and reveals the metal and its immediate environs. There are many EPR protocols with their own names. Two that have been particularly powerful for biological systems, ENDOR and ESEEM, will be discussed in this chapter in addition to continous wave (cw) EPR.

The basic phenomenon behind the three techniques – EPR (electron paramagnetic resonance), ENDOR (electron nuclear double resonance), and ESEEM (electron spin echo envelope modulation) – is the resonant absorption of microwaves by a sample in a magnetic field due to excitation of an unpaired electron from one state to a higher-energy state. In order to have such transitions it is necessary to have unpaired electrons and to have transitions that fall within an accessible energy range for a particular spectrometer. Because of the lack of unpaired electrons in the +5 oxidation state of vanadium, EPR, ENDOR, and ESEEM are not appropriate tools and NMR becomes the magnetic resonance technique of choice as discussed in the previous chapter. In the +3 oxidation state vanadium has two unpaired electrons. As is frequently observed in systems with even numbers of unpaired electrons, the energies of the EPR transitions for V^{+3} are outside the range that can be observed with the generally available spectrometers that operate at about 9 GHz (X-band). As a result, EPR is usually not a useful tool for studying the +3 oxidation state. The primary application of EPR to the study of vanadium is for the +4 oxidation state in which there is a single unpaired electron. In aqueous solution the stable +4 species is vanadyl ion, so the discussion in this chapter will focus on this ion.

The EPR spectra of vanadium in biological systems was reviewed comprehensively through 1980 (Chasteen, 1981). The information in that review will not be repeated here. Wertz and Bolton (1972) provides a more extensive introduction to EPR. This chapter will give an overview of the basic principles of continuous wave (cw) EPR, ENDOR and ESEEM techniques. The emphasis will be on recent examples of the applications of these techniques to help readers to assess how to use them in their own studies of vanadium in biological materials.

N. Dennis Chasteen (ed.), Vanadium in Biological Systems, 199–222.
© 1990 *Kluwer Acadmic Publishers. Printed in the Netherlands.*

Examples include cases in which vanadium is a naturally occuring component of the system and ones in which vanadyl ion is added as a spectroscopic probe, replacing metals such as Zn(II), Ca(II) or Mg(II) that do not give EPR signals. In this context it is interesting to note that when vanadyl ion was doped into zinc sulfate hepta-hydrate, it replaced a zinc-water unit (Bramley and Strach, 1985).

2. Electron Paramagnetic Resonance (EPR)

A. BASIC PRINCIPLES

Several characteristics of the vanadyl ion make it especially well suited for EPR spectroscopy. 1) Greater than 99% of all vanadium is ^{51}V which has a nuclear spin, I, of 7/2. The predominance of a single isotope makes it easier to interpret the EPR spectra than if there were multiple isotopes. 2) The single isotope with a nuclear spin of 7/2 is unique among paramagnetic metal ions so the nuclear hyperfine coupling is a signature for vanadium species. 3) The electron spin relaxation rate is relatively slow even at room temperature, which results in sharper EPR lines than for metals such as Fe(III) that have faster electron spin relaxation rates. 4) The unpaired electron is located in a d orbital that is oriented between the coordination axes in most vanadyl complexes so electron-nuclear couplings to coordinated ligands are generally small and do not cause extensive broadening of the EPR signals. The disadvantage of this feature is that it makes identification of coordinated ligands on the basis of the cw EPR spectra more difficult, but this problem can be overcome by the use of ENDOR and/or ESEEM.

When monomeric vanadyl ion is tumbling rapidly in fluid solution, the EPR spectrum consists of 8 lines due to interaction of the unpaired electron with the nuclear spin of the vanadium (cf. Figure 2 of Chasteen, 1981). The spacing between pairs of adjacent lines is called the hyperfine coupling, A, and values range between 60 to 100 G depending upon the groups that are coordinated to the metal (Chasteen, 1981). The integrated intensities of the 8 lines are equal, but the peak heights are unequal due to variations in the linewidths, which provides information on the degree of immobilization of the ion. As motion is slowed, the differences in the linewidths become more extreme and the spectrum approaches that of an immobilized sample. A discussion of the effects of tumbling rates on the linewidths in vanadyl EPR spectra is given in Chasteen (1981).

Immobilization of a vanadyl complex either by freezing the sample or by coordination to a molecule that cannot tumble readily results in EPR spectra similar to that shown in Figure 1. The spectrum in Figure 1 is for a lactoferrin complex in which the naturally occuring Fe(III) was replaced by vanadyl and the synergistic anion was carbonate (Eaton et al., 1988). In an immobilized sample there is a random distribution of orientations of the molecules with respect to the external magnetic field and the EPR spectrum is a composite of spectra from all of these orientations. EPR spectra are typically displayed as first derivatives of the intensity of the microwave energy that is absorbed, which emphasizes the

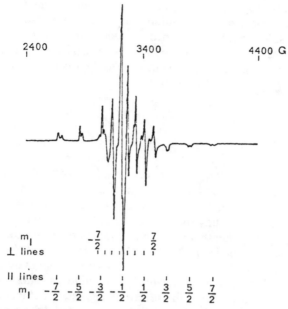

Fig. 1. 9.10 GHz EPR spectrum of vanadyl lactoferrin carbonate in water : glycerol solution obtained at 100 K with 5 G modulation amplitude and 1.3 mW microwave power. The protein concentration was about 0.5 mM and there are two vanadyl binding sites for each protein molecule. The splitting of each of the parallel lines into two peaks is due to the presence of the two vanadyl binding environments.

contributions to the spectrum from molecules with certain orientations – primarily those in which the principal axes of the vanadyl g- and A-tensors are aligned with the external magnetic field.

Vanadyl complexes typically have approximately axial symmetry with the unique axis parallel to the $V = O$ bond (Chasteen, 1981). The orientation in which this axis is parallel to the magnetic field gives the largest electron-nuclear coupling with the vanadium nucleus. The features of the EPR spectrum that arise from this orientation for each of the vanadyl nuclear spin states, for one of the vanadyl binding environments, are labelled as parallel lines in Figure 1. The orientations for which the magnetic field is in the plane perpendicular to this axis give smaller vanadium nuclear hyperfine splittings. The features for the perpendicular orientations are identified in Figure 1 as perpendicular lines.

The characteristic parameters that are obtained from the EPR spectra are the hyperfine couplings and the g values. Determination of these parameters is best done by computer simulation, but useful information can be obtained by reading values from the spectra. Close inspection of the splittings of the vanadyl signal in rapidly tumbling or immobilized environments indicates that the spacings are not equal (Chasteen, 1981; Goodman and Raynor, 1970). Although the hyperfine coupling is a constant, the details of the quantum mechanics results in terms that make the observed splittings unequal (Goodman and Raynor, 1970). These terms depend on the square of the vanadium nuclear spin state so accurate values for the hyperfine splitting can be obtained by taking the total separation between the -7/2 and + 7/2 transitions divided by 7 (Chasteen, 1981; Goodman and Raynor, 1970). Similarly the g-value can be determined approximately from the mid-point between the -7/2 and + 7/2 transitions.

B. APPLICATIONS TO BIOLOGICAL SYSTEMS

1. Calmodulin (Nieves et al., 1987)

Calmodulin is a regulatory protein that binds four Ca(II) ions. Titration of calmodulin with vanadyl ion at room temperature produced an EPR spectrum characteristic of slow tumbling, consistent with binding of vanadyl ion to the large protein molecule. This study took advantage of the fact that at pH values near physiological, vanadyl ion that was not complexed to the protein formed dimeric species that did not contribute to the EPR spectrum. Thus there was no inter- ference in the spectra from vanadyl ion that was not bound to the calmodulin. The stoichiometry indicated that the vanadyl ion bound to four sites on the protein and splittings of the parallel lines in the spectrum suggested that there were two types of binding environments. Ca(II) competes with vanadyl for binding at one type of site.

2. S-adenosylmethionine Synthetase

This enzyme catalyzes formation of a primary alkylating agent and requires two divalent metal ions (e.g. Mg(II), Mn(II), Ca(II)) and one monovalent ion (K^+) for catalytic activity (Markham, 1984). One divalent metal ion binds in the absence of substrate and the second divalent ion binds as a complex with substrate (ATP or ADP). EPR spectra as a function of vanadyl ion added showed the formation of a 1 : 1 complex with S-adenosylmethionine synthetase. The vanadyl complex did not form when both divalent binding sites were already occupied by Ca(II), which indicated that the vanadyl did not compete favorably with the Ca(II), but did bind at the same site. Spectra run in D_2O solution had narrower lines than spectra run in H_2O solution although the positions of the lines remained the same, which indicated that the structure of the vanadyl binding site remained constant (Markham, 1984). Due to differences in the nuclear magnetic moments, coupling constants to protons of coordinated water molecules are larger than for the corresponding deuterons. The reduction in the linewidths in EPR spectra for vanadyl complexes in D_2O compared with H_2O can be used to estimate the number of water molecules in the first coordination sphere (Chasteen, 1981). For this complex 1.8 $\pm$ 0.6 coordinated water molecules were calculated. Coupling (8.5 G) to the phosphorous nuclear spins of pyrophosphate (PP_i) in the complex enzyme-vanadyl-PP_i-Mg^{2+} showed that the pyrophosphate was coordinated to the vanadyl (Markham, 1984). Coordination of the pyrophosphate was also confirmed by using ^{17}O-labeled pyrophosphate and observing broadening of the vanadyl EPR signal due to interaction with the spin of the ^{17}O (I = 5/2). This series of experiments is an instructive example of the use of metal-nuclear couplings to look for coordinated groups. The large coupling constants made EPR observation of the couplings possible.

Subsequently it was shown that there was resolved hyperfine splitting between the vanadyl ion and Tl^+ (nuclear spin = 1/2) when the thallium ion bound to the monovalent binding site in the enzyme-pyrophosphate-vanadyl-Tl^+ complex (Markham and Leyh, 1987). The spectra in Figure 2 show that replacement of K^+

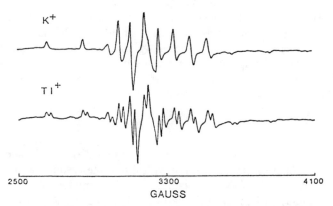

Fig. 2. EPR spectra of VO^{2+} complexes of S-adenosylmethionine synthetase with methionine and the ATP analog 5'-adenylimidodiphosphate in the presence of K^+ (A) or Tl^+ (B) at room temperature. The large protein complex tumbles at a rate that is slow compared with that required to average the parallel and perpendicular components of the EPR spectrum so the spectrum is similar to those observed for small vanadyl complexes in frozen solution. The splitting of the lines in the presence of Tl^+ is due to coupling between the vanadyl unpaired electron and the nuclear spin of the Tl. (Figure reproduced with permission of the American Chemical Society from Markham and Leyh, 1987).

by Tl^+ results in splitting of each line of the EPR spectrum into a doublet. The coupling constant was 67 MHz (24 G) and was approximately the same for the parallel and perpendicular lines of the spectrum. The isotropic coupling arises from delocalization of the vanadyl unpaired electron into the thallium orbitals that could arise from direct overlap or delocalization through a bridging ligand (Markham and Leyh, 1987). The small anisotropic contribution indicated that the vanadium-thallium distance was at least 3.6 Å.

3. Pyruvate Kinase

Pyruvate kinase binds a divalent cation and requires activation by monovalent cations (Lord and Reed, 1987). The naturally occuring monovalent cation is thought to be K^+, but other cations including Tl^+ result in some activity. In the ternary complex, enzyme-vanadyl-oxalate, the vanadyl binds at the active site and the competitive inhibitor oxalate is bound as a bidentate ligand to the vanadyl (Lord and Reed, 1987). The EPR spectrum was a typical well-resolved immobilized vanadyl spectrum with $g_{\parallel} = 1.926, g_{\perp} = 1.969, A_{\parallel} = 175 \times 10^{-4} \, cm^{-1}$ and $A_{\perp} = 67 \times 10^{-4} \, cm^{-1}$. When Tl(OAc) was added to the sample, well-resolved satellites with a splitting of 34 ± 1 G were observed both in the parallel and perpendicular regions of the spectrum. The splitting was the same at 9 and 35 GHz. The comparison of the splittings at two frequencies is an important part of this study. When the microwave frequency and magnetic field are increased by a factor of about 4, splittings that are due to g value differences increase by about a factor of 4, but hyperfine splittings are unchanged (except for the small terms that make the apparent splittings unequal). Thus the observation of the same splitting at two microwave frequencies confirms the assignment of the splitting as due to interaction of the vanadyl unpaired electron with the nuclear spin of the Tl^+. The observation of the same coupling constant in the parallel and perpendicular regions of the spectrum indicated that the coupling was isotropic and arose predominantly via through-bond interaction. This indicated that the monovalent ion bound in the vicinity of the divalent ion.

4. Bromoperoxidase

Native bromoperoxidase contains vanadium in the $+5$ oxidation state (de Boer et al., 1988a). Reduction of the enzyme with dithionite resulted in an axial EPR spectrum that was similar to what was observed for vanadyl complexes (de Boer et al., 1986, 1988a). When the pH was varied between 4.2 and 8.4, the EPR spectrum of the reduced form of the enzyme showed changes in the vanadium hyperfine splitting with little change in g values. This pH dependence was attributed to protonation of histidine and/or aspartate/glutamate residues in the vicinity of the metal. The use of D_2O and O-17 labeled water resulted in changes in the linewidths of the EPR spectrum, which indicated that there was water in the coordination sphere of the vanadium.

5. Nitrogenase

A second nitrogenase has been isolated from Azotobacter chroococum and Azotobacter vinelandii that has been shown to contain a vanadium-iron protein in place of the typical molybdenum-iron protein (Hales et al., 1987; Morningstar and Hales, 1987). The major component in the EPR spectrum has inflections with g values at 5.80 and 5.40 and was assigned to a tightly-coupled spin system with S = 3/2. The spectra of this system do not exhibit the vanadium hyperfine splitting that is characteristic of vanadyl complexes.

6. Dopamine β-monooxygenase

Dopamine β-monooxygenase is a copper-containing enzyme that catalyzes the conversion of dopamine to norepinephrine (Markossian et al., 1988). Attempts to reactivate the apoenzyme with Ni(II), Co(II), Mn(II), Zn(II), Fe(II), Fe(III), or M(VI) have been reported to be unsuccessful, but addition of excess vanadyl to the apoenzyme resulted in enzyme activity (Markossian et al., 1988). The maximal activity required a larger excess of vanadyl (350-400 fold) than was required for copper(II) (8-10 fold excess). Removal of excess vanadyl ion from the solution followed by quantitation of the EPR signal for the vanadyl enzyme complex indicated that there were 4 vanadyl ions per enzyme tetramer.

7. Vanadocytes in Tunicates

Tunicates are known to concentrate vanadium in cells that are called "vanadocytes" (Frank et al., 1986, 1988; Brand et al., 1987). Frank et al. (1986) reported that in the intact cells, most of the vanadium was in the $+3$ oxidation state, and less than 5% of the vanadium was in the $+4$ form. Fluid solution EPR spectra on solutions obtained by anaerobic lysis of vanadocytes from Ascidia ceratodes showed small concentrations of a vanadium species that was tumbling too rapidly to be coordinated to a macromolecule and therefore was assigned to vanadyl ion. A frozen sample of packed cells also gave an EPR spectrum with g and A values that were consistent with vanadyl aquo ion (Frank et al., 1986). It was argued that the linewidths of the signals and the observation of monomeric vanadyl ion were consistent only with a strongly acidic environment (estimated as pH = 1.8) in the vanadocytes.

Brand et al. (1987) challenged these reports, stating that the $+4$ vanadium signal appears only upon oxidation of the sample and had g and A values that were not consistent with the vanadyl aquo ion. The EPR spectra for the oxidized samples of Ascidia ceratodes and 11 related species gave $A_{\parallel} = 195$, $A_{\perp} = 65.0$ G, $g_{\parallel} = 1.941$, $g_{\perp} = 1.984$, while the parameters for vanadyl aquo ion are $A_{\parallel} = 202$, $A_{\perp} = 76.4$ G, $g_{\parallel} = 1.941$, $g_{\perp} = 1.985$ (Brand et al., 1987). These measurements were made on whole animals. It was also reported that when the cells from Phallusia julinea were lysed in an anaerobic environment the resulting solution was not acidic, which was in conflict with the report that the internal pH of the cell was 1.8 (Frank et al., 1986).

Frank et al. (1988) subsequently reported that their earlier results were reproducible. The frozen solution EPR spectra had $A_{\parallel} = 1.833 \times 10^{-2}$ cm^{-1}, $A_{\perp} = 0.7188 \times 10^{-2}$ cm^{-1}, $g_{\parallel} = 1.933$, and $g_{\perp} = 1.992$. They noted that the EPR measurements by Brand et al. (1987) were on whole animals and that the vanadium $+4$ environment in the tissue may be different from the intracellular environment. Anaerobic lysis of cells from Ascidia ceratodes gave solutions with pH between 1.1 and 1.4 which was consistent with their measurements of the intracellular pH based on the linewidths of the EPR signals (Frank et al., 1986). It was noted it may not be appropriate to extrapolate these results to other species, since each species may have a different blood chemistry.

The EPR spectra in blood cells of Ascidia ahodori were found to be different for samples collected from two locations in Japan (Sakurai et al., 1987). The signal from one sample indicated a stronger ligand field than was observed for the sample from the other location. These results suggest that the EPR spectrum may reflect the environment in which the organism was living.

A vanadyl EPR signal was observed for the branchial basket of *Ascidia ahodori* (Sakurai *et al.*, 1988). The branchial basket was found to have the ability to reduce vanadate to vanadyl ion, which suggests that the branchial basket may participate in vanadium accumulation from seawater.

3. Electron Nuclear Double Resonance (ENDOR)

A. BASIC PRINCIPLES

Detailed discussions of the principles and applications of ENDOR are given in Kevan and Kispert (1976) and Dorio and Freed (1979). ENDOR studies of transition metal complexes (Schweiger, 1982) and metalloproteins (Hutterman and Kappl, 1987) have been recently reviewed.

In an ENDOR experiment the magnetic field is set to a position that corresponds to a resonance in the EPR spectrum and microwave power is applied that is sufficient to partially saturate the EPR transition. The intensity of the EPR signal is recorded. A radio frequency signal is applied to the sample through a coil. The radio frequency is swept through a range of values that encompass electron-nuclear couplings. Resonant absorption of the radio frequency by the nuclear spins due to a nuclear spin resonance transition, provides a relaxation mechanism for the EPR transition. As the radio frequency passes through resonance there is an increase in the EPR signal, which is recorded as a peak in the ENDOR spectrum. Thus, ENDOR is used to measure electron-nuclear couplings.

In frozen solution or powder samples the complete EPR spectrum represents the signal for all possible orientations of the molecule with respect to the magnetic field. In an ENDOR experiment, the magnetic field is set at a particular point in the EPR spectrum, which corresponds to observation of a limited range of orientations of the molecule with respect to the magnetic field. This phenomenon is known as "orientation selection". For example, if the magnetic field is set to the low-field or high-field extreme of the EPR spectrum, only molecules with the unique axis oriented along the magnetic field are observed (Rist and Hyde, 1970). At other positions in the EPR spectrum it is necessary to account for the subset of orientations that is detected at that magnetic field (Hurst *et al.*, 1985; Hoffman and Gurbiel, 1989). In favorable cases it is possible to determine the orientation dependence of the electron-nuclear coupling constants. To first approximation the isotropic portion of the coupling is due to through-bond interaction (the contact term) and the orientation-dependent contribution is due to the through-space dipolar interaction. The use of a point-dipole approximation permits calculation of the electron-nuclear distance from the dipolar term (Hurst *et al.*, 1985; Mustafi and Makinen, 1988).

For nuclei with $I = 1/2$ such as protons, fluorine, or ^{15}N the positions of the ENDOR peaks are determined by the nuclear resonant frequency and the electron-nuclear coupling constant. Since the nuclear frequency is determined by the magnetic field at which the ENDOR experiment is done, the only adjustable parameter in analyzing the spectra is the coupling constant. For nuclei with $I >$ 1/2 such as ^{14}N, the positions of the ENDOR peaks are determined by the nuclear quadrupole coupling as well as the nuclear frequency and the electron-nuclear coupling.

There are two principal advantages of ENDOR over EPR in measurements of electron-nuclear couplings. 1) The linewidths in ENDOR spectra are narrower than the linewidths in EPR spectra, which results in improved resolution. 2) In general there are fewer lines in an ENDOR spectrum than in an EPR spectrum. In EPR spectra, coupling to n equivalent nuclei with spin, I, results in $2nI + 1$ lines. In an ENDOR spectrum coupling to n equivalent nuclei with spin 1/2 results in a pair of lines, independent of n. Somewhat more complicated ENDOR patterns are observed for $I > 1/2$ due to the presence of nuclear quadrupole couplings, but again these patterns are independent of the number of equivalent nuclei. For example, in an EPR spectrum, coupling of a metal ion to the protons of two equivalent water molecules would result in splitting of the EPR signal into 5 lines, but the ENDOR spectrum would consist of a pair of lines. This simplification is particularly important when there is interaction with multiple inequivalent nuclei. In the EPR spectrum the number of lines is equal to the product of the values of $(2nI + 1)$ for each set of inequivalent nuclei. Thus, if in addition to the pair of water molecules, the metal ion were coupled to a nitrogen ($I = 1$), the EPR signal would consist of 5x3 lines. However, the addition of a nitrogen nucleus would only add a second set of lines to the ENDOR spectrum. Since proton ENDOR lines typically occur at higher frequencies than ^{14}N ENDOR lines, the two types of nuclear interactions could be distinguished readily.

Along with these advantages come two disadvantages. 1) To be able to obtain an ENDOR spectrum it is necessary to partially saturate the EPR spectrum. For most metal ions, including vanadyl ion, it is not possible to saturate the EPR spectrum at room temperature with the microwave power available in commercial EPR spectrometers. Thus, ENDOR spectra of vanadyl complexes must be run on frozen solutions or single crystals at low temperatures. The examples cited below were obtained at temperatures between 5 and 100 K. The resolution of the ENDOR spectra is usually better at 5 K than at 100 K, due to the slower relaxation rates at lower temperature. 2) Since ENDOR spectra are not sensitive to the number of equivalent nuclei, only to the number of differently interacting nuclei, other methods must be used to determine how many nuclei of each type are in the metal coordination sphere.

B. APPLICATIONS TO VANADYL COMPLEXES

Although ENDOR holds substantial promise for studies of vanadyl complexes of biological samples, we are not aware of any published reports of its application. To give a flavor of the potential utility of ENDOR in studies of vanadyl complexes of biological materials, several studies of non-biological vanadyl chelate complexes are described below.

1. Vanadyl Tetraphenylporphyrin (Mulks and van Willigen, 1981)

^{14}N ENDOR spectra of vanadyl tetraphenylporphyrin were recorded for frozen solutions at ~ 100 K. When the magnetic field was set to the position of the lowest field peak in the EPR spectrum ($m_I = -7/2$), the ENDOR spectra were for molecules in which the V = O bond was approximately parallel to the external magnetic field. Since the vanadyl-nitrogen bonds are perpendicular to the V = O bond, the ENDOR spectra obtained at this magnetic field reflected the component of the vanadyl-nitrogen coupling constant perpendicular to the vanadyl-nitrogen bond. ENDOR data were also obtained at a magnetic field that corresponded to a perpendicular line in the EPR spectrum, which selected orientations of the molecule in which the V = O bond was perpendicular to the external magnetic field. At this magnetic field the ENDOR spectra reflected the anisotropy of the vanadyl-nitrogen coupling in the porphyrin plane. The spectra were computer-analyzed to obtain the component of the vanadyl-nitrogen coupling constant along the vanadium-nitrogen bond and perpendicular to it. The observed components of the hyperfine coupling matrix were in the range of 2.9 to 9.7 MHz. The components of the nitrogen quadrupole coupling matrix also were determined.

Proton ENDOR spectra also were obtained at magnetic fields that correspond to parallel and perpendicular orientation of the V = O bond relative to the external magnetic field. Deuterated solvents were used to eliminate ENDOR signals that arose from weak interactions of the vanadyl unpaired electron with solvent molecules. Deuteration of the phenyl rings was used to assign the signals for the pyrrole protons. The resolution of the proton ENDOR spectra was better at 15 K than at 100 K. Analysis of the orientation-dependent component of the coupling to the pyrrole protons in terms of the point dipole approximation gave a vanadium to proton distance of 5.24 Å, which is consistent with distance obtained from X-ray crystallographic studies.

2. Vanadl bis(acetylacetonate) Complexes (Kirste and van Willigen, 1982)

Proton and ^{14}N-ENDOR spectra of vanadyl bis(acetylacetonate), $VO(acac)_2$, complexes with methanol, pyridine, 2-picoline, 4-picoline, and 4-cyanopyridine were obtained in frozen solution at 100 K. As in the prior example, two magnetic fields were chosen that permitted selective observation of molecules with the V = O bond either parallel or perpendicular to the external magnetic field. The magnitudes of the hyperfine coupling constants indicated that the methanol bound trans to the V = O bond, but that the substituted pyridines could bind either cis or trans to the V = O bond. It is important to note that this conclusion could not have been obtained directly from the EPR spectra since coordination of these ligands had negligible impact on the g values and the observed changes in the A values did not appear to be significantly different for the cis and trans adducts.

3. Vanadyl Imidazole Complexes (Mulks et al., 1982)

Proton, ^{14}N- and ^{15}N- ENDOR spectra were obtained for vanadyl complexes of imidazole, histidine, and carnosine (a multidentate ligand) in frozen solution at 100 K. The ENDOR spectra were sensitive to the relative orientations of the imidazole planes in the complex. For the imidazole complex the data showed that there were four coordinated imidazoles.

4. Vanadyl Ion on Y Zeolite (van Willigen and Chandrashekar, 1983)

On the basis of the g and A values in the EPR spectra of vanadyl ion exchanged into Y zeolite it had been concluded that the principal species was $VO(H_2O)_5^{2+}$ (Martini et al., 1975). Proton ENDOR spectra recorded at 20 K were consistent with the assignment. One interesting feature was that the "matrix peak" due to weakly interacting protons in the bulk solvent was weaker than for vanadyl ion in frozen aqueous solution, which indicated that in the zeolite fewer water molecules are in close proximity to the $VO(H_2O)_5^{2+}$ than in aqueous solution. A weak interaction with a neighboring Na^+ was also detected. Dehydration of the sample resulted in loss of the proton ENDOR, which was attributed to coordination to the zeolite matrix instead of coordination to water. Upon dehydration the interaction with Na^+ become stronger, which indicated a shorter distance between the vanadyl and the sodium ion.

5. Vanadyl Ion in Methanol (Mustafi and Makinen, 1988)

This paper demonstrates the detailed stuctural information that can be obtained from ENDOR spectra. Proton ENDOR spectra of the vanadyl ion were obtained in CH_3OH, CD_3OH, and CH_3OD solutions at positions in the EPR spectra that corresponded to orientations of the V = O bond parallel and perpendicular to the external magnetic field. Figure 3 is an example of the resulting spectra. For proton ENDOR, the lines occur at the free proton frequency ± 0.5 times the electron-nuclear coupling, A. The x axis of Figure 3 is in frequency units relative to the free proton frequency. This method of labeling can be recognized because zero frequency occurs in the middle of the spectrum. This presentation facilitates comparison of spectra taken at different magnetic fields and therefore at different values of the free proton frequency. Note, however, that ENDOR spectra (including another figure in the same paper) are often presented with the x axis labeled in absolute frequency units.

Lines in Figure 3 that were present in CH_3OH solution, but absent in CH_3OD solution are due to OH protons. Analogously, the use of CD_3OH permitted assignment of the lines due to interaction with the methyl protons. In this spectrum 10 pairs of lines due to 10 different types of protons could be identified. Examination of the orientation dependence of the electron-nuclear couplings gave the isotropic and anisotropic components of the vanadyl-proton couplings. Interpretation of the anisotropic components in terms of a point-dipole model gave

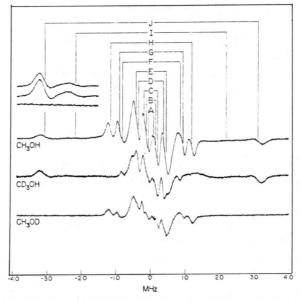

Fig. 3. Proton ENDOR spectra of the VO^{2+} ion in methanol. The magnetic field was set at the low-field end of the EPR spectrum (ca. 2800 G) which corresponds to the $m_I = -7/2$ line for molecules with the V = O bond parallel to the external field. The samples were prepared in CH_3OH, CD_3OH, and CH_3OD solutions. Spectra were recorded at 5 K. The ENDOR lines are identified in the stick diagram and occur in pairs with equal shifts to high- and low-frequency of the the free-proton frequency at 11.88 MHz. (Figure reproduced with permission of the American Chemical Society from Mustafi and Makinen, 1988).

estimates of the metal-proton distances. These distances gave a picture of the solvation structure of the complex. Information was obtained for the axial OH (trans to the V = O) in the first coordination sphere, for OH hydrogen-bonded to the V = O, for in-plane and out-of-plane OH of equatorially bound methanol, for outer-sphere OH in the equatorial plane, for methyl groups in the axial position, and for outer-sphere methyl groups in both the axial and equatorial positions. ^{13}C-CH_3OH also was used to characterize the interaction with the methyl groups. The ^{13}C-ENDOR provided values of the couplings to methyl groups in the axial and equatorial positions, and for outer-sphere methyl groups in the equatorial plane. Comparison with results from molecular modeling and X-ray crystal structures indicate that the ENDOR results underestimated the vanadium-proton distances in the equatorial plane by 0.1 to 0.3 Å, but that overall the ENDOR results provided reasonable estimates of distances.

3. Electron Spin Echo Envelope Modulation, ESEEM[1]

A. BASIC PRINCIPLES

ESEEM is a pulsed time domain EPR technique. The first question that may come to the reader's mind is, "why are pulsed techniques less common for EPR than for NMR"? There are two major factors. 1) *Timescale*. For many EPR samples, relaxation times are of the order of microseconds or less, whereas in NMR, relaxation times commonly are of the order of milliseconds or greater. As a result of the faster relaxation times, the timescale for pulsed EPR has to be faster than for NMR. Whereas pulsed NMR experiments work with pulse lengths in microseconds, pulsed EPR experiments need to work with pulse lengths in nanoseconds. The electronics needed to work on this timescale has been developed more recently than that required for pulsed NMR. 2) *Bandwidth*. High resolution pulsed NMR experiments typically work with spectral widths of the order of kHz. Since 1 G is approximately 3 MHz, it is evident that useful FT EPR spectra require much greater bandwidths than FT NMR. Since bandwidth is inversely related to pulse lengths, this requirement also contributes to the need for short pulses. The large bandwidth presents another challenge. The cavities that are used for continuous wave experiments as described in section II, are highly-tuned resonant devices with a narrow bandwidth. To do pulsed EPR it is necessary to modify the cavity to obtain a larger bandwidth or to use alternate resonators that are broader banded (Crepeau *et al.*, 1989). Current techniques permit FT EPR spectra to be taken with spectral windows of about 200 MHz (about 70 G).

The following paragraphs are intended as an introduction to the technique. More detailed discussions with references to the fundamental papers in the field are given in reviews of the applications of ESEEM to biological materials (Mims and Peisach, 1981, 1989; Tsvetkov and Dikanov, 1987).

Although ESEEM is a pulsed EPR technique, it differs from FT EPR and FT NMR in that it does not use the FID (free induction decay). Instead ESEEM, as the name implies, is based on detection of a spin echo. The two most common experiments are shown schematically in Figure 4.

In a two pulse, or primary echo, experiment a 90° pulse is used to flip the electron spin magnetization from the + z axis (in the rotating frame) into the xy plane. After a delay, τ, during which the spins tend to dephase due to differences in their resonant frequencies, a 180° pulse is applied. After this pulse, the differences in resonant frequencies which had been causing the spins to dephase, now cause the spins to refocus and form an echo, which occurs at time τ after the second pulse. Rotating frame explanations of echo formation are given in (Mims and Peisach, 1981, 1989; Tsvetkov and Dikanov, 1987). The amplitude of the echo is measured as a function of τ. If there are no nuclear spins that interact with the electron spin, the amplitude of the echo as a function of τ is an exponential decay. The time constant for the decay is called the phase memory time. For many

[1] Note: some papers use the abbreviation ESEM for the same phenomenon.

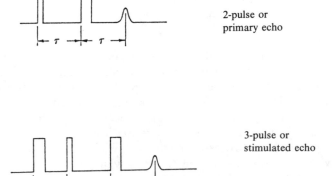

2-pulse or
primary echo

3-pulse or
stimulated echo

T is stepped

Fig. 4. Schematic representations of the pulse sequences for two-pulse (A) and three-pulse (B) electron spin echo experiments. These pulses are drawn as ideal 90° and 180° pulses. Actually any two pulses will create an echo so there are additional "unwanted" echoes in the three-pulse experiment. See Mims and Peisach (1981).

systems the phase memory time is equal to T_2. If nuclear spins interact with the unpaired electron, these interactions contribute to differences in the resonant frequencies for the nuclei. When the value of τ is varied, the amplitude of the echo depends on the cosine of the product of τ times the nuclear coupling constant in frequency units. (For explicit equations see ref. (Mims and Peisach, 1981, 1989; Tsvetkov and Dikanov, 1987). The result is a cosine function, characteristic of the nuclear interaction, superimposed on the exponential decay. Interactions with multiple nuclei result in products of the cosine functions for each of the nuclear interactions, all superimposed on the exponential decay. Isotopic substitution experiments involving, for example, ^{12}C (no nuclear spin) and ^{13}C ($I = 1/2$) can be performed by taking the ratio of data sets obtained with the two isotopes. The modulation effects of all the other nuclei cancel and the ratio contains only the ^{13}C modulation.

A three-pulse, or stimulated echo, experiment consists of a series of three pulses as sketched in Figure 4. The value of T is varied and the amplitude of the echo is measured as a function of $T + \tau$. In the absence of nuclear interaction an exponential decay is obtained with a time constant that can be as long as T_1. Interaction with nuclear spins results in a cosine-function modulation of the amplitude of the echo. Since T_1 is usually longer than T_2 in frozen solution, three-pulse echoes have the advantage that the nuclear modulation frequencies can be observed for a larger number of cycles of the cosine function, thereby giving better definition of the characteristic nuclear frequencies. These comments might give the impression that three-pulse ESEEM is the method of choice. However, the interpretation of data obtained as ratios of two sets of ESEEM data (uses are described below) is more straight-forward for two-pulse ESEEM than for three-pulse ESEEM (Tsvetkov and Dikanov, 1987). In some cases nuclear frequencies have been observed more clearly in two-pulse ESEEM than in three-pulse ESEEM (Tipton *et al.*, 1989; Eaton *et al.*, 1989).

In order to perform these experiments it must be possible to saturate the spin system, so the relaxation times of the unpaired electron must be relatively long. For vanadyl ion this necessitates the use of low temperatures. The vanadyl data discussed below were obtained at temperatures between 4.2 and 20 K. Since the echo modulation arises from different extents of refocusing of the echo, a coupling constant can only be observed if the two electron energy levels that differ by the magnitude of the coupling constant are excited simultaneously by the pulses. The maximum coupling constant that can be observed by ESEEM typically is smaller than the maximum value that can be measured by ENDOR. An anisotropic contribution to the coupling, either from a dipolar interaction or a quadrupolar interaction, is also necessary for the observation of ESEEM.

The experimental data consist of echo amplitude as a function of τ (two-pulse) or $T + \tau$ (three-pulse), which can be analyzed by measuring the periods of the cosine functions. The depth of the modulation depends on the number of equivalent nuclei and the distance between the paramagnetic center and the nuclei (Mims and Peisach, 1981, 1989; Tsvetkov and Dikanov, 1987) Visual analysis of the cosine functions becomes difficult when multiple frequencies are present. Fourier

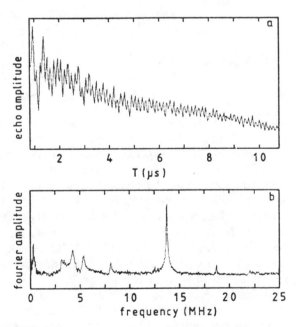

Fig. 5. Three-pulse ESEEM data (A) and the corresponding Fourier transform (B) obtained for reduced vanadium bromoperoxidase. The magnetic field was 3240 G and the microwave frequency was 9.043 GHz. The value of s was 380 ns. (Reproduced with permission of Federation of European Biochemical Societies from de Boer *et al.*, 1988b)

transforms are commonly used to convert the time-domain experimental data into the frequency domain. Since the pulses are of the order of 100 watts, the amplitude of the echo is many orders of magnitude smaller than the amplitude of the pulses. This causes a "dead-time" after the pulse during which time the cavity response to the pulse is dominant. This "gap" in the data is often called the "missing" data and can be handled in one of two ways. Time-zero for the Fourier transform can be defined to be at the start of the data and the power spectrum can be calculated. More commonly, estimates of the "missing" data are made, based on the known data, and a cosine Fourier transform is calculated. An iterative procedure (Mims, 1984) is used widely to estimate the data during the instrument dead-time. The advantage of the cosine transform is that it retains phase information that is lost in the power spectrum. Phase information is particularly helpful in interpreting 2-pulse ESEEM since peaks that arise from combinations of the fundamental frequencies are 180° out-of-phase from the peaks due to the fundamental frequencies.

Analogous to ENDOR experiments, ESEEM experiments are performed at a particular value of the external magnetic field. Thus the discussion in Section III concerning orientation selection in ENDOR spectra also applies to ESEEM.

ESEEM studies of vanadyl ion in the following small molecules have been reported: vanadyl ion in water and methanol (Dikanov et al., 1979), vanadyl bis(acetylacetonate) complexes (Astashkin et al., 1985), vanadyl complexes of NH_3 and H_2O adsorbed on silica (Narayana et al., 1985), and vanadyl complexes with coordinated nitrogens (Reijerse et al., 1989).

B. APPLICATIONS TO BIOLOGICAL SYSTEMS

1. Bromoperoxidase

As discussed above, cw EPR studies of vanadium bromoperoxidase from Ascophyllum nodosum had indicated that the reduced form of the enzyme contained vanadyl ion (de Boer et al., 1986, 1988a). Three-pulse ESEEM were obtained for a 2.3 mM sample of dithionite-reduced bromoperoxidase at 10 K at magnetic fields that correspond to a perpendicular orientation of the molecule in the magnetic field and an intermediate orientation (Figure 4a) (de Boer et al., 1988b). Signal-to-noise was inadequate to obtain data for the parallel orientation. Fourier transformation of the data gave the characteristic frequencies for the interacting nuclei (Figure 4b). The strongest peak is at 13.8 MHz, which is the proton resonant frequency at this field. The peak is due to interaction with water molecules in the vicinity of the vanadyl ion. This peak was completely replaced by one at 2.0 MHz (the deuteron resonant frequency at this field) when a sample of reduced bromoperoxidase was dissolved in D_2O. Thus the proton peak is due entirely to exchangeable protons. In addition to the proton peak, the Fourier transform spectrum (Figure 4a) also shows peaks at 3.1, 4.2, 5.3, and 8.1 MHz, which are assigned to coordinated [14]N. Similar results were obtained at a magnetic field that corresponds to a perpendicular orientation of the vanadyl ion. On the basis of comparison of the nitrogen frequencies and modulation depths with values obtained for small-molecule vanadyl complexes, it was concluded there were one or more nitrogens coordinated in the equatorial plane of the vanadyl ion.

2. Transferrin and Lactoferrin

The vanadyl ion has been used extensively to examine the metal binding sites of transferrin and lactoferrin (Chasteen, 1981). Since the naturally-occuring metal is Fe(III) and it is known to have a histidine nitrogen in its coordination sphere for both transferrin (Bailey et al., 1988) and lactoferrin (Anderson et al., 1987), it is

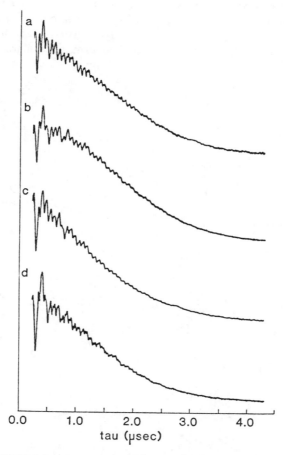

Fig. 6. Two-pulse ESEEM data for vanadyl transferrin carbonate (A), vanadyl transferrin oxalate (B), vanadyl lactoferrin carbonate (C), and vanadyl lactoferrin oxalate (D) obtained at 20 K at a magnetic field of 3250 G and a microwave frequency of 9.2 GHz. This magnetic field corresponds to a perpendicular orientation of the V = O bond in the external magnetic field. A weaker contribution from parallel orientations is also present. The protein concentration was 0.5 mM.

important to know whether other metals that are believed to bind at the same metal-binding site, also are coordinated to an imidazole nitrogen. In an ENDOR study of vanadyl substituted transferrin, there was no evidence of coupling to a coordinated nitrogen (van Willigen and Chasteen, unpublished). However, ESEEM can be more sensitive to small electron-nuclear couplings than ENDOR. In addition the relative rates of relaxation of the electron and nuclear spins can cause difficulties in ENDOR that are not present in ESEEM. Thus a negative result from an ENDOR study does not preclude a successful ESEEM result.

Two-pulse ESEEM data for vanadyl complexes of transferrin and lactoferrin with carbonate or oxalate as the synergistic anion are shown in Figure 6. The echo modulation pattern is similar for the four complexes. The highest frequency oscillation is due to protons. It is clear that one or more other frequencies are present, but it would be difficult to determine them without a Fourier transform. The Fourier transforms of the data for vanadyl transferrin oxalate and vanadyl lactoferrin oxalate are shown in Figure 7 and confirm the similarities between the

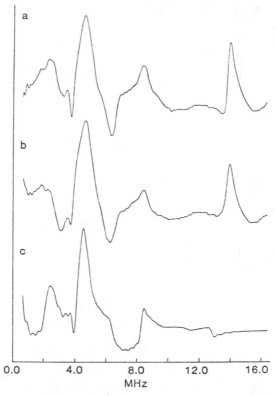

Fig. 7. Fourier transform of the two-pulse ESEEM data for vanadyl transferrin oxalate (A) and vanadyl lactoferrin oxalate (B) and a computer simulated spectrum. Details of the simulations can be found in Eaton *et al.* (1988).

two complexes. The peak at 14.0 MHz is due to protons in the vicinity of the vanadyl ion. The peaks at ~ 2, 4.8, and 8.5 MHz are due to interaction with a coordinated nitrogen. This assignment was confirmed by comparison with data for model complexes (Eaton et al., 1988). The negative-going peaks are due to frequencies that are combinations of the fundamental frequencies. For example, the negative-going peak at about 6.4 MHz is the sum of 1.6 and 4.8 MHz. A simulated spectrum is shown in Figure 6c. The details of the simulation are described in ref. (Eaton et al., 1988). In these complexes the vanadyl-nitrogen coupling constant is substantially larger than the nitrogen quadrupole coupling or the nitrogen nuclear frequency. In this limit, the simulated spectrum is more sensitive to the value of the coupling constant than to other parameters. The simulated spectrum in Figure 6c was obtained with a coupling constant of 7.0 MHz. The vanadyl-nitrogen coupling constant for the analogous carbonate complexes was 6.6 MHz. These values are consistent with coupling constants obtained by ENDOR for nitrogen coordinated to vanadyl ion in other species (see Section 3). Thus, the ESEEM data clearly show that a nitrogen is coordinated to the vanadyl ion in both the transferrin and lactoferrin complexes and that the interaction with the coordinated nitrogen is not strongly affected by the change of the anion from carbonate to oxalate.

Another important question for transferrin and lactoferrin complexes is the interaction between the metal ion and the synergistic anion. Vanadyl complexes of transferrin and lactoferrin were prepared with ^{13}C-oxalate and carbonate (Eaton et al., 1988). The two-pulse ESEEM data for the ^{13}C-labeled complexes were divided by data obtained under identical conditions for the ^{12}C-analogs. In the ratio, only the ^{13}C modulation is observed (see Figure 8 of Eaton et al., 1988). Interaction with other nuclei remains constant and does not contribute to the ratio. Modulation at the characteristic ^{13}C frequency was observed in the ratios of the data for each of the complexes. The greater depth of the ^{13}C modulation for the oxalate complexes than for the carbonate complexes suggested that in the former case there were two ^{13}C nuclei interacting with each vanadyl ion, which implied bidentate coordination of the oxalate (Eaton et al., 1988). Simulation of the depth of the ^{13}C modulation gave an estimated vanadium-carbon distance of 2.7 Å.

To examine the interaction with exchangeable protons, samples were prepared in D_2O : glycerol-d_3 (Eaton et al., 1988). Ratios of the ESEEM data obtained in D_2O to data in H_2O showed deep deuterium modulation (see Fig. 7 of Eaton et al., 1988). Analysis of the depth of the modulation and the time-dependence of the modulation depth indicated multiple water molecules within 3.4 to 4.0 Å of the metal, but there was no evidence for a directly coordinated water.

3. Pyruvate Kinase

Pyruvate kinase requires a divalent and a monovalent cation ion for activation (Lord and Reed, 1987; Tipton et al., 1989). Vanadyl can function as the divalent ion, replacing the naturally occurring Mg^{2+}.

In Fourier transforms of 3-pulse ESEEM data obtained for a vanadyl-pyruvate

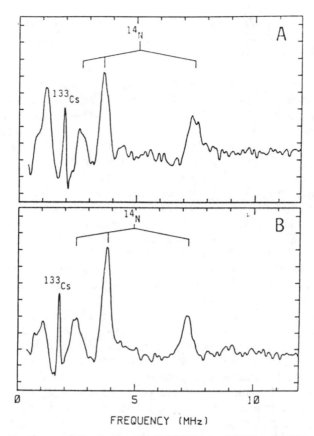

Fig. 8. Fourier transforms of the three-pulse ESEEM data for (A) vanadyl-pyruvate kinase · Cs⁺-pyruvate and (B) vanadyl-pyruvate kinase · Cs⁺-phosphoenolpyruvate obtained at 4.2 K and a microwave frequency of 8.8 GHz. (Reproduced with permission of the American Chemical Society from Tipton *et al.*, 1989).

kinase-phosphoenolpyruvate complex at 8.8 GHz (3130 G) and 10.0 GHz (3870 G), peaks were observed that did not scale linearly with magnetic field (Tipton *et al.*, 1989). This behavior is characteristic of ^{14}N peaks (and peaks for other I > 1/2 nuclei with significant quadrupole couplings), but not for ^{13}P (or other nuclei with I = 1/2). Thus the peaks were assigned to ^{14}N. This is an important example of the use of multiple microwave frequencies to assist in the assignment of ESEEM signals. By analogy with frequencies observed for model complexes, the nitrogen was assigned to lysine rather than a histidine. The vanadyl-nitrogen coupling constant was found to be 4.9 MHz.

Fourier transforms of two-pulse ESEEM data for vanadyl-pyruvate kinase-[1-^{13}C]pyruvate at 11.4 GHz and 4060 G showed peaks at 3.4 and 5.6 MHz (Tipton *et al.*, 1989). The ^{13}C nuclear frequency at this magnetic field is 2.2 MHz. Thus the two frequencies observed represent a pair of peaks centered at the free carbon frequency with a splitting of 2.2 MHz, due to vanadyl-carbon coupling. A coupling of this magnitude indicates through-bond interaction and shows that the pyruvate was coordinated to the metal. It had been shown previously (Lord and Reed, 1987) that oxalate binds to the divalent metal.

In Fourier transforms of the three-pulse ESEEM data for vanadyl-pyruvate kinase-oxalate complexes, signals were observed at the nuclear frequencies for the monovalent cations Na^+ and K^+ (Tipton et al., 1989) which indicated that the monovalent cations were in close proximity to the vanadyl ion. Since the peaks were not shifted from the nuclear resonance frequencies, there was negligible through-bond interaction between the vanadyl ion and the monovalent cations. In the vanadyl-pyruvate kinase complexes with pyruvate and phosphoenol-pyruvate, interaction with Na^+ and Cs^+ was observed. The data for the Cs^+ complexes are shown in Figure 8. The peak for the Cs^+ is readily observed in addition to the peaks for the coordinated nitrogen. Note that the interaction with the coordinated nitrogen was largely unchanged by the changes in the cation or coordinated ligand.

These three ESEEM studies of vanadyl bound to proteins have appeared within the year prior to the writing of this article. The insights provided by these studies suggest that ESEEM will be a powerful method for studying other vanadyl complexes.

5. Summary

EPR spectroscopy can provide information about vanadyl ion in biological systems. The g- and A- values can be used to characterize the vanadyl binding environment. Quantitation of the EPR signal can be used to determine the number of moles of coordinated vanadyl ion. CW EPR linewidths reflect the rate of molecular tumbling. Coupling of the vanadyl unpaired electron to nuclear spins of phosphate and Tl^+ has been observed, but coupling to coordinated nitrogen is generally too small to be resolved in CW EPR spectra. Increases in CW EPR linewidths in the presence of D_2O or O-17 labeled water indicate water coordination to the vanadium. ENDOR and ESEEM hold considerable promise for detailed characterization of the nuclear spins in the vicinity of the vanadyl ion. Although there are regions of overlap in which either technique can be used, larger coupling constants are more readily observed by ENDOR and smaller coupling constants are more readily observed by ESEEM. Coordinated water molecules can be studied in much greater detail with ENDOR than by CW EPR. Both ENDOR and ESEEM have been used to study nitrogens in the vanadyl coordination sphere.

Acknowledgment

The work in our laboratory on vanadyl complexes of transferrin and lactoferrin was supported by NIH Grant GM21156. We thank Dr. G. D. Markham for providing comments on the manuscript.

References

Anderson, B. F., Baker, H. M., Dodson, E. J., Norris, G. E., Rumball, S. V., Waters, J. M., Baker, E. N.: 1987, Structure of Human Lactoferrin at 3.2 – A Resolution, *Proc. Natl. Acad. Sci. (U. S.)* **84**, 1769-1773.

Astashkin, A. V., Dikanov, S. A., Tsvetkov, Yu. D.: 1985, Coordination of Vanadyl Acetylacetonate with Nitrogen-Containing Donor Bases, *Zh. Strukt. Khim.* **26**, 53-58 (pp. 363-368 in transl.).

Bailey, S., Evans, R. W., Garratt, R. C., Gorinsky, B., Hasnain, S., Horsburgh, C., Jhoti, H., Lindley, P. F., Mydin, A., Sarra, R., Watson, J. L.: 1988, Molecular Structure of Serum Transferrin at 3.3 – A Resolution, *Biochemistry* **27**, 5804-5812.

Bramley, R., Strach, S. J.: 1985, Reexamination of the Location of VO^{2+} by EPR and ZFR, *J. Chem. Phys.* **82**, 2437-2439.

Brand, S. G., Hawkins, C. J., Parry, D. L.: 1987, Acidity and Vanadium Coordination in Vanadocytes, *Inorg. Chem.* **26**, 626-628.

Chasteen, N. D.: 1981, 'Vanadyl(IV) EPR Spin Probes, Inorganic and Biochemical Aspects', in L. J. Berliner and J. Reuben, eds., *Biological Magnetic Resonance* 3, Plenum Press, New York, pp. 53-119.

Crepeau, R. H., Dulcic, A., Gorcester, J., Saarinen, T. R., Freed, J. H.: 1989, Composite Pulses in Time-Domain ESR, *J. Magn. Reson.* **84**, 184-190.

de Boer, E., van Kooyk, Y., Tromp, M. G. M., Plat, H., Wever, R.: 1986, Bromoperoxidase from *Ascophyllum nodosum*: a Novel Class of Enzymes Containing Vanadium as a Prosthetic Group?, *Biochim. Biophys. Acta* **869**, 48-53.

de Boer, E., Boon, K., Wever, R.: 1988a, Electron Paramagnetic Resonance Studies on Conformational States and Metal Ion Exchange Properties of Vanadium Bromoperoxidase, *Biochemistry* **27**, 1629-1635.

de Boer, E., Keijzers, C. P., Klaassen, A. A. K., Reijerse, E. J., Collison, D., Garner, C. D., Wever, R.: 1988b, ^{14}N-Coordination to VO^{2+} in Reduced Vanadium Bromoperoxidase, an Electron Spin Echo Study, *FEBS Lett.* **235**, 93-97.

Dikanov, S. A., Yudanov, V. F., Tsvetkov, Yu. D.: 1979, Electron Spin-Echo Studies of Weak Hyperfine Interactions with Ligands in some VO^{2+} Complexes in Frozen Glassy Solution, *J. Magn. Reson.* **34**, 631-645.

Dorio, M. M., and Freed, J. H. (eds.): 1979, *Multiple Electron Resonance Spectroscopy*, Plenum, New York.

Eaton, S. S., Dubach, J., More, K. M., Eaton, G. R., Thurman, G., Ambruso, D. R.: 1988, Comparison of the Electron Spin Echo Envelope Modulation (ESEEM) for Human Lactoferrin and Transferrin Complexes of Copper(II) and Vanadyl Ion, *J. Biol. Chem.* **264**, 4776-4781.

Eaton, S. S., Dubach, J., Eaton, G. R., Thurman, G., Ambruso, D. R.: 1989, Electron Spin Echo Envelope Modulation (ESEEM) Evidence for Carbonate Binding to Iron(III) and Copper(II) Transferrin and Lactoferrin, submitted for publication.

Frank, P., Carlson, R. M. K., Hodgson, K. O.: 1986, Vanadyl Ion EPR as a Noninvasive Probe of pH in Intact Vanadocytes from *Ascidia ceratodes*, *Inorg. Chem.* **25**, 470-478.

Frank, P., Carlson, R. M. K., Hodgson, K. O.: 1988, Further Investigation of the Status of the Acidity and Vanadium in the Blood Cells of *Ascidia ceratodes*, *Inorg. Chem.* **27**, 118-122.

Goodman, B. A. and Raynor, J. B.: 1970, Electron Spin Resonance of Transition Metal Complexes, *Adv. Inorg. Chem. Radiochem.* **13**, 135-362.

Hales, B. J., Morningstar, J. E., Case, E. E.: 1987, ESR Spectroscopic Properties of Vanadium-Containing Nitrogenase from Azotobacter Vinlandii, *Rec. Trav. Chim. Pays-Bas* **106**, 1-15.

Hoffman, B. M., Gurbiel, R. J.: 1989, Polycrystalline ENDOR patterns from Centers with Axial EPR

Spectra. General Formulas and Simple Analytic Expressions for Deriving Geometric Information from Dipolar Couplings, *J. Magn. Reson.* **82**, 309-317.

Hurst, G. C., Henderson, T. A., Kreilick, R. W.: 1985, Angle-Selected ENDOR Spectroscopy. 1. Theoretical Interpretation of ENDOR Shifts from Randomly Oriented Transition Metal Complexes, *J. Am. Chem. Soc.* **107**, 7294-7299.

Huttermann, J., and Kappl, R.: 1987, ENDOR: Probing the Coordination Environment in Metalloproteins, *Metal Ions in Biol. Syst.* **22**, 1-80.

Kevan, L. and Kispert, L. D.: 1976, *Electron Spin Double Resonance*, Wiley, New York.

Kirste, B., van Willigen, H.: 1982, Electron Nuclear Double Resonance Study of Bis(acetylacetonato)oxovanadium(IV) and Some of Its Adducts in Frozen Solution, *J. Phys. Chem.* **86**, 2743-2749.

Lord, K. A., Reed, G. H.: 1987, Vanadyl(IV)-Thallium(I)-205,203 Superhyperfine Coupling in Complexes with Pyruvate Kinase, *Inorg. Chem.* **26**, 1464-1466.

Markham, G. D.: 1984, Structure of the Divalent Metal Ion Activator Binding Site of S-Adenosylmethionine Synthetase Studied by Vanadyl(IV) Electron Paramagnetic Resonance, *Biochemistry* **23**, 470-478.

Markham, G. D., Leyh, T. S.: 1987, Superhyperfine Coupling Between Metal Ions at the Active Site of S-Adenosylmethionine Synthetase, *J. Am. Chem. Soc.* **109**, 599-600.

Markossian, K. A., Paitian, N. A., Nalbandyan, R. M.: 1988, The Reactivation of apodopamine β-monooxygenase by vanadyl ions, *FEBS Lett.* **238**, 401-404.

Martini, G., Ottaviani, M. F., Seravilli, G. L.: 1975, Electron Spin Resonance Study of Vanadyl Complexes Adsorbed on Synthetic Zeolites, *J. Phys. Chem.* **79**, 4232-4235.

Mims, W. B.: 1984, Elimination of the Dead-Time Artifact in Electron Spin-Echo Envelope Spectra, *J. Magn. Reson.* **59**, 291-306.

Mims, W. B., Peisach, J.: 1981, Electron Spin Echo Spectroscopy and the Study of Metalloproteins, *Biol. Magn. Reson.* **3**, 213-263.

Mims, W. B., Peisach, J.: 1989, ESEEM and LEFE of Metalloproteins and Model Compounds, in "Advanced EPR in Biology and Biochemistry", A. Hoff, ed., Elsevier Publishing, Amsterdam, in press.

Morningstar, J. E., Hales, B. J.: 1987, Electron Paramagnetic Resonance Study of the Vanadium-Iron Protein of Nitrogenase from Azotobacter Vinelandii, *J. Am. Chem. Soc.* **107**, 6854-6855.

Mulks, C. F., van Willigen, H.: 1981, ENDOR Study of Oxovanadium(IV) Tetraphenylporphyrin Randomly Oriented in Rigid Solution, *J. Phys. Chem.* **85**, 1220-1224.

Mulks, C. F., Kirste, B., van Willigen, H.: 1982, ENDOR Study of VO^{2+}-Imidazole Complexes in Frozen Aqueous Solution, *J. Am. Chem. Soc.* **104**, 5906-5911.

Mustafi, D., Makinen, M. W.: 1988, ENDOR-Determined Solvation Structure of VO^{2+} in Frozen Solution, *Inorg. Chem.* **27**, 3360-3368.

Narayana, M., Narasimhan, C. S., Kevan, L.: 1985, Characterization of Silica-supported Vanadium Species, *J. Chem. Soc., Faraday Trans I* **81**, 137-141.

Nieves, J., Kim, L., Puett, D., Echegoyen, L., Benabe, J., Maldonado-Martinez, M.: 1987, Electron Spin Resonance of Calmodulin-Vanadyl Complexes, *Biochem.* **26**, 4523-4527.

Reijerse, E. J., Shane, J., de Boer, E., Collison, D.: 1989, ESEEM of Nitrogen Coordinated Oxo-Vanadium(IV) Complexes in "Proceedings of the International Workshop on Electron Magnetic Resonance of Disordered Systems", Bulgaria, 1989, Scientific Publishing, Singapore, in press.

Rist, G. H., Hyde, J. S.: 1970, Ligand ENDOR of Metal Complexes in Powders, *J. Chem. Phys.* **52**, 4633-4643.

Sakurai, H., Hirata, J., Michibata, H.: 1987, EPR Characterization of Vanadyl Species in Blood Cells of Ascidians, *Biochem. Biophys. Res. Commun.* **149**, 411-416.

Sakurai, H., Hirata, J., Michibata, H.: 1988, ESR Spectra of Vanadyl Ion Detected in Branchial Basket of an an Ascidian, *Ascidia ahodori, Inorg. Chim. Acta* **152**, 177-180.

Schweiger, A.: 1982, Electron Nuclear Double Resonance of Transition Metal Complexes with Organic Ligands, *Struct. Bond.* **51**, 1-128.

Tipton, P. A., McCracken, J., Cornelius, J. B., Peisach, J.: 1989, Electron Spin Echo Envelope Modulation Studies of Pyruvate Kinase Active-Site Complexes, *Biochemistry* **28**, 5720-5728.

Tsvetkov, Yu. D., Dikanov, S. A.: 1987, Electron Spin Echo: Applications to Biological Systems, *Metal Ions Biol. Syst.* **22**, 207-263.

van Willigen, H., Chandrashekar, T. K.: 1983, ENDOR Study of VO^{2+} Adsorbed on Y Zeolite, *J. Am. Chem. Soc.* **103**, 4232-4235.

van Willigen, H., Chasteen, N. D.: unpublished work.

Wertz, J. E., Bolton, J. R.: 1972, "Electron Spin Resonance: Elementary Theory and Practical Applications", McGraw-Hill, New York.

Index